SCHRIFTENREIHE DES IMT 17

Schriftenreihe des Instituts für
Management und Tourismus

Herausgegeben von
Christian Eilzer und Bernd Eisenstein

Michael Lück / Claire Liu (eds.)

A kaleidoscope of tourism research:

Insights from the International Competence Network of
Tourism Research and Education (ICNT)

**Bibliographic Information published by the
Deutsche Nationalbibliothek**
The Deutsche Nationalbibliothek lists this publication in the Deutsche
Nationalbibliografie; detailed bibliographic data is available online at
http://dnb.d-nb.de.

Library of Congress Cataloging-in-Publication Data
A CIP catalog record for this book has been applied for at the
Library of Congress.

ISSN 2194-0002
ISBN 978-3-631-84696-4 (Print)
E-ISBN 978-3-631-85283-5 (E-PDF)
E-ISBN 978-3-631-85284-2 (EPUB)
E-ISBN 978-3-631-85285-9 (MOBI)
DOI 10.3726/b18323

© Peter Lang GmbH
Internationaler Verlag der Wissenschaften
Berlin 2021
All rights reserved.

Peter Lang – Berlin · Bern · Bruxelles · New York ·
Oxford · Warszawa · Wien

All parts of this publication are protected by copyright. Any
utilisation outside the strict limits of the copyright law, without
the permission of the publisher, is forbidden and liable to
prosecution. This applies in particular to reproductions,
translations, microfilming, and storage and processing in
electronic retrieval systems.

This publication has been peer reviewed.

www.peterlang.com

We dedicate this book to
Melville Saayman

Contents

IN MEMORIAM 9

Editorial Introduction 13

Alisha Ali and Philip Murray
Learning in the Workplace: An Innovative Approach to Work Experience .. 19

Heike Schänzel and Donna O'Donnell
School-led Tourism in New Zealand: Educating Students for Global
Citizenship 33

Elricke Botha, Petrus van der Merwe and Melville Saayman
Interpretation in Protected Areas: The case of the Kgalagadi
Transfrontier Park, South Africa 53

Chantal D. Pagel, Matthias Waltert, Michael Scheer and Michael Lück
Swimming with wild orcas in Norway: Killer whale behaviours addressed
towards snorkelers and divers in an unregulated whale watching market 77

Yasmine M. Elmahdy, Mark B. Orams and Michael Lück
Towards more sustainable marine mammal tourism in New Zealand –
reviewing the literature and identifying the gaps 99

Eva Holmberg
It is mostly about money- discussions related to the political decision-
making resulting in giant pandas moving into Ähtäri Zoo 133

Anne Köchling
Tourist Relevance and Perception of UNESCO World Heritage Sites and
National Parks in Germany: The Case Study of the "Wadden Sea" 157

Amber Knowsley, Tomas Pernecky and Jill Poulston
Motivations and Perceptions of Carbon Neutral Accommodation in New
Zealand 189

Sabrina Seeler, Michael Lück and Heike Schänzel
Increased travel experience and its effects on responsible and sustainable travel behaviour 215

Marit Gundersen Engeset and Jan Velvin
Identifying travel motives for visitors to Hemsedal during low season 239

Jarmo Ritalahti
Customer insights of sustainability and responsibility in leisure travel intermediation in Finland 253

Anne Köchling, Julian Reif and Rebekka Weis
Tourism and Political Crises: An Analysis of British Holiday Planning in Times of "Brexit" 271

List of Contributors 289

IN MEMORIAM

Professor Melville Saayman

24/05/1965 – 14/03/2019

Peet van der Merwe and Elmarie Slabbert

Professor Melville Saayman was appointed at the previously known Potchefstroom University of Christian Higher Education on the 1st of January 1992 as a lecturer in the Department of Recreation. He was instrumental in the development of tourism management as a Science offered at the University in the Faculty of Economic and Management Sciences. Melville conceptualised the development of the Tourism Programme, which involved a variety of BA, BCom and BSc packages, appropriate curriculum and textbooks to fill this gap in academia. He headed the tourism programme for 17 years (2009) after which his focus shifted to the further development of the research unit. Melville took over the Institute for Leisure Studies in 1995 and changed the name to the Institute for Tourism and Leisure Studies. In the early 2000s, the National Research Foundation (NRF) put out a call for the establishment of research entities and this led to the birth of the first tourism niche research entity called SEIT (Socio-Economic Impact of Tourism), after which the name was changed to TREES (Tourism Research in Economic, Environs and Society). In 2015, TREES became a research unit and to date is one of the most successful tourism research units in South Africa.

On a National level Melville served on several boards as a director, including the South African Tourism Board (SATOUR), North-West Parks and Tourism

Board, Institute of Environment and Recreation Management, National Zoological Council, South African National Recreation Council (SANREC), North-West Recreation Council (PROREC-NW), North West Development Corporation and Aardklop National Arts Festival. At an international level, he was a member of the executive committee of the Association of International Experts in Tourism (AIEST) and also served on the World Tourism Organisation's panel of experts. He also formed part of the ICNT network, which is a network between various universities worldwide, focusing on teaching and research in tourism. He served on various journal editorial boards and has published in most of the major national and international tourism journals. He became the first South African to be nominated as resource editor of the leading tourism journal, *Annals of Tourism Research*.

Melville was active in the field of tourism and leisure economics and development with a clear focus on poverty alleviation. Lately, he has contributed significantly to research in the fields of event economies, nature-based tourism, marine tourism, and how these types of tourism alleviate poverty. He became the first National Research Foundation (NRF) rated researcher in tourism in South Africa. From his pen, numerous leisure and tourism books (20), scientific articles (240), technical reports (420) and in-service training manuals (8) have been published. He was study leader and promoter to 98 Master's and PhD students and presented more than 100 papers at international conferences. He has also been an examiner of more than 50 Master's and PhD theses. In 2010, 2011, 2014 and 2015 Melville was awarded researcher of the year of North-West University. In 2017 he was acknowledged by the Department of Science and Technology for his contribution to Sustainable Tourism. He was also ninth on the top 100 list of Tourism scholars based on citation records compiled by Scopus (in total, only three are from Africa). Melville received a well-deserved Lifetime Achievement Award from the Tourism Educators of South Africa in 2018.

Besides his excellent career, we as a tourism family had the privilege to work closely with him, learn from him and be part of several projects, discussions and developments. He believed that in order to teach tourism effectively, you need to travel and experience what you teach first hand. We, therefore, visited many places together and built excellent industry networks through this approach. Melville did not work alone, which created opportunities for each tourism staff member to develop to the best of their ability. He had a good sense of humour, and we will not forget all the special moments we shared with a lot of laughs and good times. He always offered advice, had a plan and believed that we can do whatever we put our minds to – this empowered us to believe in what we do. Melville changed lives, gave direction and positively did this. He lived his life to the fullest and the memories created through this will always be with us. Melville will be missed by everyone that crossed his path but his dreams for the School of Tourism Management and TREES will be achieved by those he taught.

Editorial Introduction

Since 2007, ICNT partners from all corners of the world have met annually at one of the member university's home, presented and discussed a wide array of tourism issues, and explored the local areas, sampling tourism products. This group of dedicated tourism researchers is as diverse as a tourism group can be, but they have one thing in common: A passion for the tourism and hospitality industry, sustainable development and operations, teaching and learning, and a love for people. This diversity is fascinating, and regularly results in publications that are as diverse as the authors. This is the 6th volume stemming from the work of ICNT partners, which is published as part of the overall book series of the Institute for Management and Tourism (IMT) (renamed in September 2020 to German Institute for Tourism Research) at the West Coast University for Applied Sciences in Heide, Germany. Previous volumes covered topics, such as national park tourism (Eilzer et al., 2008), global tourism experiences (Eilzer et al, 2011), social aspects of tourism (Lück et al, 2015), perspectives on destination management and tourist experiences (Lück et al., 2016), and multi-stakeholder perspectives in tourism (Ali & Hull, 2018). This volume once again showcases a kaleidoscope of tourism and hospitality topics, ranging from tourism education, to sustainable tourism, wildlife tourism, crises in tourism, and to travel intermediation, and tourist motivation and experiences.

Universities are constantly being challenged to innovate to keep ahead of the dynamic and volatile operating environment. In the first chapter, Ali and Murray introduce the reader to Sheffield Hallam University's response to the increasing pressure graduates experience during their job search post university: They introduce an innovative way of learning by partnering with a large hotel in the USA, where students spend one semester and gain valuable experience as part of the Learning in the Workplace course.

In Chapter 2, Schänzel and O'Donnell report on school-led overseas trips as an endeavour to create global citizens. Employing an interpretive paradigm, this exploratory study provides insights into the socio-cultural experiences and meanings gained by nine New Zealand students on an international school trip to Thailand.

Moving from formal education in the previous two chapters to informal visitor education in the South African Kgalagadi Transfrontier Park, Botha and colleagues provide an analysis of interpretation and, through a case study, provide practical guidelines on how interpretation can be incorporated practically in protected areas. Their study identifies the interpretation needs tourists expect from this park. Not only are specific interpretation services identified, but the tourists also show particular interest in specific topics (e.g., climatology or ecology), as well as preferences in specific media in which such interpretation should be presented.

The demand for closer and more personal interactions with marine wildlife has been catered for by the commercialisation of in-water interactions with various species of marine wildlife. Norway is the only country in the world where commercial and unregulated swim-with-programmes with orcas have been regularly offered on a small scale. Orcas are highly mobile predators and, where potentially dangerous animals are involved, humans are at risk of experiencing aggressive, defensive or dangerous play behaviour. In Chapter 4, Pagel and colleagues report on a baseline study in Norway, investigating the long-term effects of human–wildlife interactions with wild orca. The present results facilitate the general interpretation of interspecific killer whale behaviour, which is essential for safer interactions with these marine predators and enables a sound comparison with similar studies featuring other odontocetes.

Staying with marine wildlife, Elmahdy and colleagues, in the following chapter, review pertinent literature to identify gaps in the quest to achieve sustainable marine mammal tourism in Aotearoa New Zealand. Their findings illustrate that there is a need for an investigation and assessment of the effectiveness of the current marine mammal tourism management regime in New Zealand to ensure the long-term sustainability of the industry and conservation of the targeted species.

In Chapter 6, Holmberg takes the reader to the Finnish Ähtäri Zoo, which has hosted giant pandas as a loan from China since 2018. Analysing published newspaper and Internet articles, Holmberg explores how the controversial political decisions related to accepting the pandas to Finland and Ähtäri Zoo were discussed in Finnish media.

Köchling introduces the reader to the German National Park Wadden Sea, and how visitors perceive this park. She examines the general interest of German residents in a journey to UNESCO world heritage sites and national parks, the impact of the designations UNESCO world heritage site and national park on German residents' destination choice, and the perception of all German UNESCO world heritage sites and national parks as well as four international reference destinations by German residents. The study also provides implications for the sustainable development of national parks.

In Chapter 8, Knowsly and colleagues, present a case study in New Zealand which aimed to identify the motivations and initiatives involved in curbing carbon emissions in the accommodation sector. The only carbon neutral hotel in New Zealand, Sudima Hotel Auckland Airport was selected as the sample. Data analysis from the hotel management and the customers shows that ecological responsibility and competitiveness were key motivating factors. The findings in this study also provide practical value to businesses interested in implementing carbon-mitigating initiatives while maintaining customer satisfaction.

Focusing on the experienced and hybrid tourists, Seeler and colleagues (Chapter 9) present the findings of a mixed-methods study conducted in Germany and New Zealand to explore the relationship of travel experiences and sustainable tourist behaviours. The quantitative survey results demonstrate that tourists with higher experience levels tend to participate in more sustainable forms of travel and experienced tourists show high interest for activities and experiences that positively contribute to the sustainable development of the tourism industry.

In Chapter 10, Engeset and Velvin present the findings of a visitor survey on the motivations, satisfaction, and future visit intentions of visiting Hemsedal, Norway. This study identifies several different types of push and pull motivations for tourists visiting Hemsedal during the summer season. This study provides implications for the development of destinations with high seasonal fluctuation in demand.

Using quantitative methods, Ritalahti (Chapter 11) analyses the importance of sustainability and responsibilities of travel agencies from customers' perspectives. This Finnish study illustrates the challenges to traditional travel intermediation businesses. The participants of the research have emphasized the importance of responsible operation and the values of looking after the wellbeing of its employees.

In the final chapter, Köchling and colleagues examine whether the UK leaving the European Union (Brexit) would affect British travel plans in 2017 in general and their intention to visit Germany as a destination. An online panel survey with 2000 people was conducted in UK. The results show that Brexit seems to have little impact on the planned holiday travel behaviour of the British travelling to Germany, and their interest in participating in domestic tourism in the UK remains the same.

With the wide range of interest of the individual authors, this volume once again provides a wide range of tourism research around the globe. We would like to take the opportunity to thank all authors for their generous contributions to this book: Alisha Ali, Elricke Botha, Yasmine Elmahdy, Marit Engeset, Eva Holmberg, Amber Knowsley, Anne Köchling, Philip Murray, Donna O'Donnell, Mark Orams, Chantal Pagel, Tomas Pernecky, Jill Poulston, Julian Reif, Jarmo Ritalahti, Melville Saayman, Heike Schänzel, Michael Scheer, Sabrina Seeler, Elmarie Slabbert, Peet van der Merwe, Jan Velvin, Matthias Waltert, and Rebekka Weis. Thank you also to Bernd Eisenstein, who is the master behind the scenes, and to Christian Eilzer for his admirable and meticulous work to always get our ICNT books to the publisher!

It is with great sadness, that we learned about the passing of one of our core ICNT members, and dear friend, Melville Saayman. We have all seen him in high spirits at the 2018 meeting, which he co-hosted at North-West University in Potchefstroom, South Africa. Melville was always the sunshine of the ICNT meetings, in good spirits, full of humour and energy. He was gentle and kind to presenting colleagues and (sometimes very nervous) postgraduate students, asking pertinent questions without intimidation. One could bet that he would ask a question about "memorable experiences", something he was passionate about, and he certainly left us with many memorable experiences from over ten ICNT conferences. ICNT is not the same without Melville, and he will be sorely missed. But his spirit lives on, as he has taught us so much about compassion and laughter, which we will carry on for many years to come. We dedicate this book to his memory, and his closest colleagues and friends, Peet and Elmarie, provided us with a memoriam at the beginning of this volume.

Michael Lück and Claire Liu
Auckland, New Zealand

References

Ali, A., & Hull, J. S. (Eds.). 2018. *Multi-stakeholder perspectives of the tourism experience: Responses from the International Competence Network of Tourism Research and Education (ICNT)*. Peter Lang Verlag.

Eilzer, C., Arlt, W.G., & Eisenstein, B.(Eds.).(2011). *Global experiences in tourism: Proceedings of the International Competence Network of Tourism Management (ICNT)*. Martin Meidenbauer Verlag.

Eilzer, C., Eisenstein, B., & Arlt, W.G.(Eds.).(2008). *National parks and tourism: Answers to a global question from the International Competence Network of Tourism Management (ICNT)*. Martin Meidenbauer Verlag.

Lück, M., Ritalahti, J., & Scherer, A. (Eds.). (2016). *International perspectives on destination management and tourist experiences. Insights from the International Competence Network of Tourism Research and Education (ICNT)*. Peter Lang Verlag.

Lück, M. Velvin, J., & Eisenstein, B. (Eds.). (2015). *The social side of tourism: The interface between tourism, society, and the environment. Answers to global questions from the International Competence Network of Tourism Research and Education (ICNT)*. Peter Lang Verlag.

Alisha Ali and Philip Murray

Learning in the Workplace: An Innovative Approach to Work Experience

1. Introduction

Universities are constantly being challenged to innovate to keep ahead of the dynamic and volatile operating environment. To add value to employability and to society in a wider context, educators must have a thorough understanding of key stakeholder needs, particularly employers, to inform and enrich the curriculum (Barber, Deale, & Goodman, 2011). This paper presents an innovative, flexible, and sustainable higher education model designed and delivered at Sheffield Hallam University (SHU), *Learning in the Workplace*. Students spend a semester of their second-year learning and working with an employer. The employer serves as multi-disciplinary learning laboratory where students engage in active experimentation working and learning alongside experienced personnel in a range of roles aligned to their curriculum and degree choice. Rather than completing their modules in the classroom at SHU, the students on this experience develop their learning through hands-on work experience with the employer.

High enrolment growth at universities in western societies, like the United Kingdom, leads to intense competition in the graduate labour market. Whilst the perceived employment of hospitality students increases during their course, their confidence in actually gaining employment decreases (Beaumont, Geyde, & Richardson, 2016). To better enhance the employability of these hospitality graduates, there is a need for more practically orientated education and strengthening of university-industry collaborations (Guermat, Saad, & Boutifour, 2015; Pani, Das, & Sharma, 2015; Wang & Tsai, 2014).

The new approach presented in this chapter is a solution to this dilemma as it presents an innovative way of partnering with industry. This learning experience meets the needs of and draws on the strengths of business, academia and learners. SHU's ambition is to become the UK's leading applied University and it continues to break down the traditional boundaries between the classroom and the workplace. This approach is also a response to key stakeholder demands such as the government, professional bodies and SHU's business partners. The UK Department for Business Innovation & Skills in 2016 called for evidence related to Accelerated Courses in Higher Education. On February 24th, 2017, the UK Government announced plans for two-year accelerated degrees. An article in The Caterer (14 October 2016) discussed the need to think differently about how we deliver Hospitality Business Management education and grow future talent.

Furthermore, the neoliberal paradigm shift in academic education indicates that universities should stress less on how to intellectually enlighten graduates but rather develop employable graduates, who are enabled to gain a footstep in the highly competitive industry after their university education (Ali, Murphy, & Nadkarni, 2016; McCowan, 2015). Research in this area focusses on identifying and scrutinising specific skill sets and competencies which are demanded from university graduates in general (Finch, Hamilton, Baldwin, & Zehner, 2013; Scott, 2014) and more specifically, which factors affect the employability of hospitality graduates, either by establishing "generic attributes" (Moolman & Wilkinson, 2014), the hospitality students pespective (Beaumont, Geyde, & Richardson, 2016), or analysing the perspective of hospitality employers and managers (Alhelalat & Al-Hussein, 2015; Ali, Murphy, & Nadkarni, 2016) on employability.

The Learning in the Workplace programme at SHU is an example of a hybrid work experience provision which has characteristics from both the concept of work-based learning (WBL) as mode and field of delivery conceptualisation (Costley & Armsby, 2007). Credit is awarded for assessment based on a single semester of work with the partner organisation. The programme learning outcomes are generic and address the skills and the competencies required by hospitality employers. The classroom is the partner's property with university tutors facilitating student learning and providing general support for the practice-based knowledge which is developed by students.

This learning experience offers not only an example of best practice in curriculum design and industry partnership but also potential insights into two further key areas in hospitality education research similar to those identified by Kim and Jeong (2018); the use of technology to augment experiential learning experiences for students and the development of graduates' cultural intelligence as they are exposed to international experiential learning environments.

2. Literature Review

Learning in the workplace has been the focus of considerable academic interest for almost 30 years and has evolved over this period through two notable phases. The first was a focus on pedagogy and the second concentrated on how to operationalise WBL with a focus on stakeholder partnerships, assessment and student experience. This literature review discusses the key issues of pedagogy, and partnership with a specific focus on the hospitality literature. The WBL is applied to all learning that occurs within the workplace or arises directly out of workplace interests and frequently overlaps with experiential learning, informal learning and continuous professional development (Lester & Costley, 2010).

A great deal of the WBL literature can be viewed as having an underlying epistemological position, often implicit, relating to whether WBL is a mode or field of study. Costley and Armsby (2007) assert that this difference is essentially whether a degree is granted through WBL with generic assessment criteria (field of study), or by WBL with subject specific criteria (mode of study) which are typically associated with work placements or undergraduate 'sandwich courses'. The hospitality literature is clearly oriented towards WBL as a mode of study with limited contributions in pedagogy (Gruman, Barrows & Reavley, 2009). However there has been a strong focus on the development of partnerships between higher education institutions and employers.

It has been universally acknowledged that experiential learning should be an essential component in any hospitality curriculum. This method of learning through experiences has been proven to be instrumental in stimulating students' interest in the subject under study, improving their satisfaction and fosters the development of a range of personal, professional and technical skills (Snyder, 2003; Kong & Yan, 2014). This approach necessities a strong connection between industry and academia. Moreover, industry partners place a premium on the competencies, skills and abilities acquired through internships and placement (Petrillose & Montgomery, 1997; Smith & Cooper, 2000; Yiu & Law, 2012). The partnerships between universities and employers feature strongly in the WBL literature. Such partnerships and the depth of the relationship are found to be very influential in the success of UK hospitality programmes (Busby & Gibson, 2010). The partnerships between higher education institutions and employers are critical in delivering flexible, accessible and responsive hospitality curricula and providing graduates with the skills and competencies to be successful in the evolving global economy (Sewanger & Gursoy, 2007; Ferrandez-Berrueco, Kekale & Devins, 2016).

Little consensus exists on the design of hospitality curricula and the core competencies which are required by graduates (see Leung, Wen, & Jiang, 2018 for a comprehensive review of international approaches) and the role required from industry in curricula design (Ali, Murphy & Nadkarni, 2016). Employer involvement in the WBL process has been defined as arm's length; where the university is merely a provider of labour and partnerships, the university and employer seek to develop a mutually beneficial understanding with recruitment fluctuating dependant on employer needs, or strategic alliances where the employer works with a single university in the main and commits to continued recruitment of students over the period of the relationship (Cassells, 1994, cited in Busby & Gibson, 2010, p. 6). Undoubtedly, providing a meaningful and beneficial WBL experience requires a 'unique tripartite relationship between hospitality students, employers, and educators' (Yiu & Law, 2012, p. 392). We believe that the Learning in the Workplace experience described in this chapter is an example of such a strategic relationship and provides students with a rewarding workplace practice.

Moore (2007) asserts that the university-employer relationship requires clear vision and rigorous frameworks in place to support independent learning to meet the needs of both parties. Reeve and Gallacher (2005) in their call for further research into the forms and effectiveness of sustained partnerships, suggest that universities need to reflect more fully on the nature of the contribution they can make to the WBL process.

While there is broad agreement about the definition of WBL, there is considerable discussion around how universities define their WBL provision (Reeve & Gallacher 2005), with a significant variability in how WBL is operationalised across university programmes and how deeply it is ingrained in curricula. The Quality Assurance Agency for Higher Education (2019, p.6) subject benchmark statements for events, hospitality, leisure, sport and tourism states 'courses in hospitality have evolved significantly beyond this vocational focus to combine technical, management and scientific disciplines as a coherent whole, while retaining strong connections to industry'. However, Leung, Wen, and Jiang's (2018) analysis of international hospitality curricula concludes that UK hospitality programmes are more business orientated without any specific hospitality industry focus. These contrary views would suggest that there is a need for new and innovative programmes that work directly with industry to deliver real world skills to graduates and there remains space for institutions to innovate in their curriculum design and provision. Most of the writing on hospitality WBL concentrates on work placement or internship but does not discuss structured learning initiatives as identified in the Learning in the Workplace experience.

Criticisms of WBL are infrequent in the academic literature but voiced more informally by stakeholders (Costley & Armsby, 2007). Some of the concerns noted by Lester and Costley (2010) pertain to the lack of substantive curriculum, formal assessments or examinations, the transition of tutors from teacher to facilitator engendered by the WBL process and an associated drop in quality related to the commercialisation of WBL programmes to increase student numbers. These authors further assert that these challenges are not different from what 'can be expected when any new paradigm or method of organising emerges' (p. 569). Our approach provides details on a unique partnership and WBL provision within the hospitality field and demonstrates how these barriers can be overcome.

3. The Learning in the Workplace Experience

The development of Learning in the Workplace allows SHU to offer choice and flexibility to the 'new' generation of learners based on their individual circumstances and career paths. This framework allows SHU's students to spend a semester of their second year of study in the workplace and achieve 60 credits toward their final degree classification. This length of time in the workplace will be viewed as a placement and allows students the accelerated opportunity to complete their studies with work experience in three rather than four years. By successfully completing this pathway, students will receive the additional 'placement' in brackets in their degree titles. Should students desire to complete a further 48 weeks work experience, this opportunity will also be available to them. Students learn through their job roles, tasks and projects related to learning outcomes with guidance from tutors at SHU and workplace supervisors. Each student will be expected to meet agreed job objectives, meet learning outcomes for the Learning in the Workplace module (see below) and reflect on and articulate their learning.

The Framework

For SHU's students the focus will be on 'learning while working, learning how to do new or existing things better, learning that takes place in the workplace and a 'curriculum' that grows out of the experience of the learner, their work context and their community of practice' (SHU Work Based Learning Framework, p.6).

Learning in the Workplace has the following characteristics:

- Require students to relate theory to practice through critical reflection and from a personal and professional perspective.
- Enable students to take responsibility for their own learning, including its nature, and within certain parameters, its focus and pace.
- Develop innovative learning, teaching and assessment strategies appropriate to the workplace.

- Support the professional and personal development of our students.
- Recognise workplace learning as legitimate. Upon successful completing of the work experience and assignments students will be awarded 60 credits of study.
- Allows programme teams the flexibility of implementing Learning in the Workplace in either Semester 1 or 2 of the second year of study or both semesters of the second year of study.
- Students will be able to obtain 'placement' in brackets in their degree titles on successful completion of this pathway.
- Delivery may be adapted to meet the requirements of a placement provider and students. For example, if students are going to the USA to undertake their Learning in the Workplace they will have to complete a minimum of 32 hours of work per week to be able to meet the visa regulations.

The learning and teaching strategy is designed to engage students actively with their acquisition of the skills required of future business leaders. The module enables students to build upon triggers of learning from their work, independent study time, seminars, workshops or professional reviews. Students will be provided with academic support to understand the fusion between the theory and practice to achieve the learning outcomes. It is generally accepted that students learn most effectively when working at their own pace and in their own time (Tullis & Benjamin, 2011).

Learning in the Workplace at Business A

The framework discussed above was implemented in cooperation with Business A which embodies an entrepreneurial spirit and when fused with disciplined marketing and management experience, leads to creative solutions. Business A was actively involved in the development of this experience and were very responsive partners. Working collaboratively, SHU and Business A designed a cutting-edge Learning in the Workplace programme where students studying for an undergraduate degree in hospitality, tourism or events can spend either semester 1 or 2 of their second-year living, earning and learning at Business A. When the students were on site at Business A, they undertook two rotations each lasting two months and aligned to their degree. The learning outcomes of the existing curriculum were mapped against the indicative job roles. For example, events students completed an events rotation.

Students were given synchronised time off for study which was usually a Tuesday. This enabled students to have the time to meet up for group work and facilitated a virtual classroom with tutors in Sheffield. Support mechanisms were built into the programme for students as listed below. To ensure a great learning experience, students were provided support before, during and after their time at Business A.

Support Prior to Arriving at Business A

Students are supported with the following:

- Visa support if required.
- Support in helping them to re-let accommodation already committed to in Sheffield.
- Cultural and academic briefings prior to departure.
- A winter/summer induction school. This provided an opportunity for the cohort to form an identity, have an introduction to their academic studies and prepare for their work and cultural experiences. During this time, they were introduced to their tutors, the modules, and the assessments. It also allowed them some time to familiarise themselves with this new approach to learning. Some key lectures relating to the modules were also delivered during this time.
- Students completed a health and safety test prior to departure to Business A.
- All other support mechanisms which are already available to students if they were on-site.

Support whilst at Business A

- Induction was on the first day with introduction to the site and programme.
- All students had coordinated time off per week for study and to facilitate group work.
- Monthly excursions and guest talks.
- Academic and pastoral support was provided face-to face and virtually.
- A member of the academic team visited close to assessment submission time to provide support.
- Briefings were held with staff at Business A to prepare them for welcoming the SHU cohort and for their roles in supporting the students during the WBL experience.
- Support for students seeking a further placement was provided virtually.
- Stable Wi-Fi and a SHU classroom.
- Reading materials accessible online.
- All other support mechanisms which are already available to SHU students were accessible to these students.

Support on return to SHU

- Reflective session on how their learning has developed their personal, professional and technical skills.

The Module

The module begins with a comprehensive induction/orientation which took place prior to departure. Induction sessions not only provide useful advice and information for students on their module delivery and tasks, but also cover the practical, cultural, and legal aspects of living, working and studying in their host country. There is also a strong focus on assuring the students that the support systems of the university are still available and accessible if required. As the programme has evolved the induction was expanded to reflect the experience of the students and now includes advice on issues such as managing homesickness and being assertive in the job role. Fundamentally, the induction serves as a team building role as many participants from differing course routes are meeting and working together for the first time.

Students were provided with comprehensive preparation materials which outline their roles and the requirement of self-motivation and management for successful completion of the module. The managers at Business A were also provided with briefing materials outlining their roles as mentors and guides for the students during the programme. The employer was involved in the definition of tasks and projects to be completed by the students. The summative assessment included an analysis of their progress and outcomes in working towards and achieving these activities, and employers were invited to attend the presentation portion of the assessment package and to ask questions and offer feedback in tandem with the academics.

The Learning in the Workplace module is divided into three assessment tasks to ensure the experience is manageable for students and to provide milestones to keep them engaged with their academic work. The assessments require students to investigate their workplace, to source, analyse and discuss industry and academic relevant information and to examine what they have achieved in the workplace, and how strengths have been built upon, using appropriate theories on reflection and personal development.

4. The Success

Learning in the Workplace is anchored in the philosophy of 'praxis', where in this context theory and practice integrate and emerge in the workplace. SHU and Business A were very committed to developing the nexus between the classroom and the workplace. This has been the key to success as both parties share a common vision and are always looking for creative solutions to develop future talent. This programme has also been effective because student expectations are carefully managed. Constant communication with the employer has allowed the SHU team to hone the emphasis on specific aspects of the rotations the students complete and work more closely with Business A on providing relevant roles for the students, so they see the value to their degree and their future careers. To ensure the principles of the programme were adhered to and to ensure its success, investment in the programme required the following:

- Dedicated academic lead and account manager working closely with Business A management, students, academics and student support.
- Modification of the existing curriculum and ensuring the academic learning outcomes are met through the Learning in the Workplace Programme.
- Training of the staff at Business A and academics at SHU to ensure they were aware of the expectations of the programmes and how to support students.
- Working with the students to prepare them for the new mode of work, study, cultural awareness and health and safety.
- Site visits to the employer and visits to SHU by the employer.

Student feedback has been overwhelmingly positive. They were also achieving better grades because of this experience, on average an increase of 3–5 %. Some students' comments are as follows:

> This allowed me to gain a greater understanding of how the hotel operated as an entity whereas before I had never worked in a hotel which will be extremely beneficial to my career. The program also was extremely beneficial for me academically. In my first semester (whilst in the UK) I averaged a 57.75 % for my modules, however for my second semester (whilst working in the US) my grade average was 71.25 % which is a 13.5 % increase as seen in the figure below. I feel that the knowledge and skills that I was learning allowed me to apply them directly to my studies. Furthermore, as I am studying International Hospitality Business Management, the opportunity has given me international experience in which I was able to work in a different culture.

> I learnt a wide range of skills that will help me in my career, but it also allowed me to grow as a person. The rotational aspect of the work allowed me to work in 5 different departments where I was trained to a supervisor level.

> During the internship, I was promoted to a supervisor which has become a big stepping stone for my career. Not only by promoting me, but by also mentoring me along the way, I was able to develop managerial skills which will also support me in my academic work in my final year.

This innovation has also provided benefits for the employer as depicted in their comments below:

> *The SHU students bring to the organisation a heightened sense of commitment and enthusiasm, which is contagious among the rest of our staff. These students have evidenced a strong passion for hospitality and the various roles within the industry, as well as the strong contribution each makes to the overall effort of serving guests and creating outstanding memories for them. In addition, they have been eager to share with fellow employees and guests alike their own cultures, and to learn and experience what they can, which has brought a marvelous sense of joie de vivre to the organisation. We are very pleased to be able to work with these exceptional students, and very proud to be able to collaborate with Sheffield Hallam University.*

The success of this innovative learning experience has led to external recognition as it was nominated for two national awards and successfully won one in innovation in teaching and learning.

5. Conclusion

Education, like any other sector, must strive to innovate. In hospitality we can do so by forming meaningful partnerships with employers to maximise the benefit for our students and provide them with the skills to become agile and resilient business leaders. Work experiences are critically important in achieving this. However, for this to add value, such practices must reflect the changing needs of our students and our industry. We must therefore think creatively and seek new ways of delivering valued practical experiences beyond the traditional models. The above case describes an example of how this can be achieved. The aim of the experience is to amalgamate practical learning within the existing curriculum for the development of the next generation of business leaders. It sought to provide students with an international learning experience through working and studying in a different culture. The success of this programme is rooted in the commitment of both the employer and SHU. Such partnerships must be carefully harnessed and invested in to develop maximum value for all in building a sustainable education model for the future.

Future development would revolve around continuing to learn from the process and being open to evolving the induction and the module itself based on feedback from all parties. Further work will be undertaken in widening this programme to other employers, not only in hospitality but in other sectors. The onus should not only be on higher education providers to create such opportunities, but employers should be innovative and creative and take a risk on such types of learning as the long-term benefits accrued can be limitless for their businesses.

As this programme is relatively new, future research will focus on the value of virtual learning and the how this can develop graduates' cultural intelligence. Hospitality is international, and graduates should be able to work across boundaries and cultures. An experience, such as Learning in the Workplace will strengthen their cultural quotient and their ability to be adaptable and flexible. Additionally, research also needs to be conducted on investigating students' deep learning during these experiences and the relationship to their attainment gap at university and career success.

References

Alhelalat, A.J., & Al-Hussein, B. (2015). Hospitality and non-hospitality skills between education and industry. *Journal of Business Studies Quarterly*, 6(4), 46–55.

Ali, A., Murphy, H., & Nadkarni, S. (2016). Hospitality employers' perceptions of technology for sustainable development: The implications for graduate employability. *Tourism and Hospitality Research*, 18(2), 131–142.

Barber, N., Deale, C., & Goodman, R. (2011). Environmental sustainability in the hospitality management curriculum: Perspectives from three groups of stakeholders. *Journal of Hospitality and Tourism Education*, 23(1), 6–17.

Beaumont, E., Geyde, S., & Richardson, S. (2016). Am I employable?': Understanding students' employability confidence and their perceived barriers to gaining employment. *Journal of Hospitality, Leisure, Sport & Tourism Education*, 19, 1–9.

Brodie, P., & Irving, K. (2007). Assessment in work-based learning: investigating a pedagogical approach to enhance student learning. *Assessment & Evaluation in Higher Education*, 32(1), 11–19.

Busby, G. D., & Gibson, P. (2010) Tourism and hospitality internship experiences overseas: A British perspective. *Journal of Hospitality, Leisure, Sport & Tourism Education*, 9(1), 4–12.

Costley, C., & Armsby, P. (2007) Work-based learning assessed as a field or a mode of study. *Assessment & Evaluation in Higher Education*, 32(1), 21–33.

Ferrández-Berrueco, R., Kekale, T., & Devins, D. (2016). A framework for work-based learning: basic pillars and the interactions between them. *Higher Education, Skills and Work-Based Learning*, 6(1), 35–54.

Finch, D. J., Hamilton, L. K., Baldwin, R., & Zehner, M. (2013). An exploratory study of factors affecting undergraduate employability. *Education + Training*, 55(7), 681–704.

Gruman, J., Barrows, C., & Reavley, M. (2009). A hospitality management education model: Recommendations for the effective use of work-based learning in undergraduate managent courses. *Journal of Hospitality & Tourism Education*, 21(4), 26–33.

Guermat, C., Saad, M., & Boutifour, Z. (2015). Special issue: University–industry interaction and new role of universities. *International Journal of Technology Management and Sustainable Development*, 14(2), 67–69.

Kim, H. J., & Jeong, M. (2018). Research on hospitality and tourism education: Now and future. *Tourism Management Perspectives*, 25(October 2017), 119–122.

Kong, H.Y., & Yan, Q. (2014). The relationship between learning satisfaction and career competencies. *International Journal of Hospitality Management*, 41, 133–139.

Lester, S., & Costley, C. (2010). Work-based learning at higher education level: Value, practice and critique. *Studies in Higher Education*, 35(5), 561–575.

Leung, X. Y., Wen, H., & Jiang, L. (2018). What do hospitality undergraduates learn in different countries? An international comparison of curriculum. *Journal of Hospitality, Leisure, Sport & Tourism Education*, 22, 31–41.

McCowan, T. (2015). Should universities promote employability? *Theory and Reserach in Education*, 13(3), 267–285.

Moolman, H. J., & Wilkinson, A. (2014). Essential generic attributes for enhancing the employability of hospitality management graduates. *Tourism: An International Interdisciplinary Journal*, 62(3), 257–276.

Moore, L. J. (2007). Partnerships and work-based learning: An evaluation of an opportunity to pioneer new ways to care for the older people in the community. *Assessment & Evaluation in Higher Education*, 32(1), 61–77.

Pani, A., Das, B., & Sharma, M. (2015). Changing Dynamics of Hospitality & Tourism Education and its Impact on Employability. *Parikalpana: K I I T Journal of Management*, 11(1), 1–12.

Petrillose, M. J., & Montgomery, R. (1997). An exploratory study of internship practices in hospitality education and industry's perception of the importance of internships in hospitality curriculum. *Journal of Hospitality & Tourism Education*, 9(4), 46–51.

Quality Assurance Agency (2019). Subject benchmark statement: Events, Hospitality, Leisure,

Sport and Tourism *Retrieved August 18, 2020 from* https://www.qaa.ac.uk/docs/qaa/subject-benchmark-statements/subject-benchmark-statement-events-leisure-sport-tourism.pdf?sfvrsn=c339c881_11

Reeve, F., & Gallacher, J. (2005). Employer-university 'partnerships': A key problem for work- based learning programmes? *Journal of Education and Work*, *18*(2), 219–233.

Scott, B. (2014). Graduate attributes and talent perceptions: Reflections of the first year of graduate employment. *International Journal of Employment*, *22*(1), 39–59.

SHU (2015). Work Based Learning Framework. Retrieved April 4, 2017 from https://portal.shu.ac.uk/sites/aqf/cda/pages/wblf.aspx

Smith, G., & Cooper, C. (2000). Competitive approaches to tourism and hospitality curriculum design. *Journal of Travel Research*, *39*(1), 90–95.

Snyder, K., (2003). Ropes, poles and space–active learning in business education. *Active Learning in Higher Education*, *4*(2), 159–167.

Swanger, N., & Gursoy, D. (2007). An industry-driven model of hospitality curriculum for programs housed in accredited colleges of business: program learning outcomes-part III. *Journal of Hospitality & Tourism Education*, 19(2), 14–22.

The Caterer (2016). *Shoulder to shoulder: David Foskett and Peter Jones*. Available from: https://www.thecaterer.com/articles/493607/shoulder-to-shoulder-david-foskett-and-peter-jones

Tullis, G.J., & Benjamin, S.A. (2011). On the effectiveness of self-paced learning. *Journal of Memory and Language*, *64*(2), 109–118.

Wang, Y.F., & Tsai, C.T. (2014). Employability of hospitality graduates: Student and industry perspectives. *Journal of Hospitality & Tourism Education*, *26*(3), 125–135.

Yiu, M. & Law, R. (2012). A review of hospitality internship: different perspectives of students, employers, and educators. *Journal of Teaching in Travel and Tourism*, 12(4), 377–402.

Heike Schänzel and Donna O'Donnell

School-led Tourism in New Zealand: Educating Students for Global Citizenship

1. Introduction

For many decades, school children have actively engaged in educational experiences outside the classroom. In recent years, school-led tourism in (Western) developed countries has expanded to include international school trips, with around 97 % of those secondary schools providing opportunities for school students to travel overseas (Campbell-Price, 2014; Ritchie, 2009; Xplore Camps, 2017). These trips traditionally focus on curriculum-related foreign languages and are organised as opportunities to learn and interact with other cultures (Byrnes, 2001). However, international school-led tourism increasingly includes a range of extra-curricular activities with the overall aim of educating school students for global citizenship (Tarrant et al., 2014). This reflects a shift away from conventional concepts of citizenship based on promoting nationalism and patriotism, towards educating (young) people on how "to deal with the complex issues and interconnectedness of life in a highly globalised world" (Hermann, Meijer, & Van Koesveld, 2016, p. 132). Considerable future growth is predicted for these types of school-led trips.

The meanings of holidays for children and adolescents and the role of travelling in constructed subjectivities are largely absent in academic research (Schänzel & Smith, 2014; Small, 2008). Even less is known about educational tourism, and particularly international school excursion tourism (Campbell-Price, 2014; Cooper & Latham, 1988; Larsen & Jenssen, 2004). As future tourists, the children's views of their holiday experiences are significant (Cullingford, 1995). Adolescence (puberty to age 19) is of interest here, as travel-related competencies, cultural beliefs, personal habits and much of what is later taken for granted are established in this phase of life (Frändberg, 2010). Short-term study trips abroad can be perceived as creative, engaging and transformative learning experiences, but require further research (Perry, Stoner, & Tarrant, 2012). The potential for cross-cultural learning is particularly high when there is extensive interaction with local people and elements of culture shock (van 't Klooster, van Wijk, Go, & van Rekom, 2008) or culture confusion (Hottola, 2004). This is especially marked when Westerners are exposed to a non-Western culture for the first time and when the educational trip is organised by the hosts, as is the case for the trip discussed in this chapter. There are also health and safety considerations that need to be considered when children travel to developing countries (Richter, 2005) that can have an influence on the educational experience.

Educational overseas trips are increasingly being offered through secondary schools in New Zealand, but not much is known about the meanings and experiences gained by the participating adolescents (Campbell-Price, 2014). In the past, research more broadly in the social sciences has been conducted *on* adolescents and has neglected the voices of the young people themselves (Bassett, Beagan, Ristovski-Slijepcevic, & Chapman, 2008). More recently, there has been a move towards research *with* adolescents, engaging them as active subjects rather than objects, recognising their rights and giving them a voice (Greene & Hogan, 2005; Grover, 2004) but tourism research still lags behind this trend. This study is based on nine secondary school students (aged 16–18 years) who took part in a three-week educational trip to Chiang Mai in Thailand in January 2013. The students volunteered to represent their New Zealand secondary school (a well-respected state school in a middle-class suburb in Auckland) and to teach English at a private college in Chiang Mai, which exposed them to varied cultural and educational experiences related to Thai customs. The trip was termed 'Language Immersion Programme (LIP) – Inbound – Thai Culture Immersion' and was organised with the overall intention, as supported by the New Zealand education agenda, of preparing students to become global citizens. For most of the students, this was their first overseas travel experience to a developing country. The aim of this chapter is to provide insights into the socio-cultural experiences and meanings gained by students from New Zealand on an international school-led trip.

2. Literature Review: Educational Travel and School-led Tourism

Since 1994, the Ministry of Education has worked in conjunction with organisations to encourage educational experiences outside the classroom for New Zealand schools and students (J. Rivers, 2006). Whilst research is limited in educational tourism, different perspectives have been examined. Several authors have adopted a tourism perspective, evaluating the economic importance of the school sector to the tourism industry (Dale, Ritchie, & Keating, 2012; Ritchie, 2009). The educational literature has generally sought to justify the benefits and effectiveness of these programmes (Ballantyne & Packer, 2002; Griffin, 1998; Price & Hein, 1991). In particular, the educational benefits of domestic school trips as part of the curriculum have featured in educational debates and discussions (Campbell-Price, 2014). However, for many students, secondary school is a time when they wish to broaden their horizons. Increased emphasis on global awareness and global exposure by society and the education system means more students are setting their sights not only on domestic tourism locations, but also on international destinations. Consequently, demand for international school-led tourism is currently expanding, with almost all secondary schools, globally, providing opportunities for their students to travel overseas (Campbell-Price, 2014; Xplore Camps, 2017). Many of these overseas opportunities include volunteer tourism.

Volunteer tourism is generally defined as travel to a destination to take part in charity projects that aid the well-being of the local community (Wearing, 2001). However, the expense of these international school visits has been criticised in the media and is proving to be an area of concern (Fox, 2011; Ibbotson, 2010), since volunteer tourism is now a billion dollar industry with volunteers often paying relatively more than they would have paid for a 'normal' package holiday to the same destination (Kennedy, 1994; Wearing, 2001). The issues relating to the high cost and the media scepticism of these ventures has led to concerns regarding their educational relevance and placed educational tourism at the forefront of education debate within New Zealand.

Nevertheless, academic discourse from the disciplines of both tourism and education acknowledges that out-of-classroom education makes learning relevant by giving authentic, hands-on experiences that support classroom learning by bringing the curriculum to life (Ministry of Education, 2014). From an educational perspective, most research has justified student participation in overseas field trips (Brown, 2005; Chen & Chen, 2011; Cohen, 2014; Wakeford, 2013). International field trips have been linked with many positive attributes including the ability to motivate individuals, develop positive traits and behaviour in adolescents, provide opportunities to build confidence, improve self-awareness, enhance cultural awareness and create a sense of belonging.

Indeed, research suggests that the desire for personal development is a primary motive for participating in educational/volunteer tourism (Chen & Chen, 2011), since travel provides opportunities to re-invent oneself or increase self-esteem through informal learning. Consequently, research conducted on overseas study tours appears to highlight the application of personal skills and knowledge as a means of motivating individuals. For example, many international language schools provide overseas travel experiences as a way of encouraging individuals to learn a foreign language. Immersing students in a language environment and exposing them to the local surrounds creates unique personal opportunities for cognitive learning that cannot be duplicated in the classroom (Carr & Cooper, 2003; Houser, Brannstrom, Quiring, & Lemmons, 2011). Thus, the learning-centred approaches created by overseas travel help motivate and engage students.

The educational literature describes a student-centred approach as 'experiential learning' (Kolb, 1984). Experiential learning is "learning in which the learner is directly in touch with the realities being studied" (Kolb, 1984, p. xviii), an ongoing process of learning directly from personal life experiences. Gilbertson (2006, p. 9) adds that "experiential learning, is learning that occurs through an authentic experience". Experiential learning stimulates the feelings and cognitive aspects of personality; a process of change resulting from formal or informal experiences that are evaluated and reflected upon to find meaning (Beard & Wilson, 2006; Curzon, 1997; Petty, 1998; Taras & Gonzalez-Perez, 2014). Research suggests that experimental learning opportunities can be utilised to develop traits and behaviour in adolescents (Ballantyne & Packer, 2002; Bodger, 1998; Scoffham & Barnes, 2011; Wakeford, 2013). For this reason, experiential learning is appropriate to adolescent school students in preparation for career development or further education and is often integrated into overseas school excursions."

Consequently, extensive research has been conducted on the personal and social responsibility development associated with overseas trips; for example, Saitow (2009) claims that educational travel provides the opportunity to build confidence, improve self-awareness and transform relationships within a group. Wakeford (2013) believed that the students themselves embraced the opportunity to make new friends by meeting and interacting with both the host community and other students within their travel group. This aligns with Larsen and Jenssen's (2004) research, which suggested the social experience gained through travel is more important than what the individuals did. However, Cushner (2004) believes that, for travel experiences to be meaningful, students need to be immersed and engaged with the local communities. He refers to these experiences as 'subjective cultural experiences', the invisible aspects of people such as attitudes, values, and expectations that form the basis for deep and meaningful learning. Whilst S. Brown (2005) points out that brief cultural encounters with local people have often proved to be the highlight of an individual's holiday.

Research often highlights changes in individual attitudes towards different cultures, improved environmental responsibility and the positive psychological changes in participants. For example, Scoffham and Barnes (2011) found learning outside the classroom evoked feelings of happiness that impacted on individual lives and learning, thus creating a sense of well-being and improving the emotional state of the individual involved. It is therefore evident from research that these educational experiences develop positive attitudes and improve behaviour and maturity. In relation to volunteer tourism, Wearing (2001) suggests that volunteer tourism has positive influences on participants and the potential to induce change. Bodger (1998) discusses this in greater depth, finding that daily exposure to a different set of cultural values encouraged a change in personal attitudes and cultural values, increasing respect for different cultures. Furthermore, meaningful interactions and connecting with others are an important element in educational tourism, enhancing cultural awareness and a sense of belonging, proving educational tourism has numerous social benefits not related to the curriculum.

Awareness of and respect for others is now considered an essential element in secondary education, with many New Zealand government agendas and policies reflecting the notion of global citizenship and internationalism (Ministry of Education, 2007a). The National Curriculum, for example, provides a framework to ensure adolescents are equipped with the skills, values and knowledge to adapt and thrive within a multicultural society, embedding global citizenship into learning experiences within the classroom and education outside the classroom (Ministry of Education, 2007b). This may explain the increase in demand for overseas trips that are both curriculum and non-curriculum related. In addition to New Zealanders travelling overseas, the International Education Agenda (Ministry of Education, 2007a) promises international student enrichment and integration into New Zealand's educational establishments, increasing global knowledge, understanding and respect for other cultures by creating meaningful interactions between international and local students.

To evaluate the relevance and importance of overseas trips, academic discourse has considered both the positive and negative aspects of the overseas experience and the cultural effects on students. Discussions in academic discourse are generally dominated by the benefits of international study tours as a means of justifying the travel experiences (Campbell-Price, 2014). Many research projects have focused on the positive benefits and learning opportunities for students (Ballantyne & Packer, 2002; Falk & Dierking, 1997; Forsey, Broomhall, & Davis, 2012; Nunan, 2005; O'Callaghan, 2006; Saitow, 2009; Wakeford, 2013). For example, O'Callaghan (2006) found that after overseas travel students viewed their own country from a different perspective and developed an appreciation for their own national identity. Cavanagh (2012) found that, post study tour, student attitudes towards education became more positive, more appreciative of the British educational system and students consequently worked harder. Thus, in these cases the students gained insights into their own culture by stepping outside it and looking in, hence changing their understanding and worldviews. This is substantiated in other studies (Nunan, 2005; Williams, 2005), which focused on intercultural communication, understanding and sensitivity. For example, Williams (2005) noted that the meaning of the study tour experience is often characterised by cultural empathy, as perceptions and attitudes change as intercultural communication skills are developed. Nevertheless, Coleman (1997) claims the effects of studying overseas varies considerably from one individual to another and are dependent on the individual's own culture and upbringing. Whilst many disadvantages of overseas trips, such as language difficulties (Ferrante, 2013), loneliness, isolation (Eckert, Luqmani, Newell, Quraeshi, & Wagner, 2013) and culture shock (Cushner, 2004), are acknowledged in educational literature, these negative experiences are often discussed as providing valuable learning experiences for the individuals involved and are ultimately still perceived as beneficial. After all, intercultural experiences can be a unique learning process leading to individual growth and development. Hence, educational tourism can be a useful tool for the development of life skills that support identity and social development.

As volunteer tourism gains momentum and increases in popularity, research continues to expand with many academics seeking to determine if the length of stay influences the effect on learning. Research examining travel and programme duration has started to emerge and much controversial debate surrounds this issue. The majority of these studies compared the effects of long-term and short-term participation on personal learning; Landis and Brislin (2013), for example, found no significant differences. A view shared by Paige, Fry, Stallman, Jon and Josic (2010) who found that it did not matter where the students went or for how long, it was the quality of the programme that instigated and generated learning. In addition, McMillian and Opem (2002) also suggest the length of stay is unimportant, and it is the impact of the experience that affects the individuals for the rest of their life. However, Dwyer (2004) found that 'more is better' and argued the longer students are abroad the more significant the learning; this is certainly well documented in studies that focus on language acquisition (W. Rivers, 2008). In contrast, Fischer (2009) adopts a different stance, suggesting that short-term study trips abroad can be just as beneficial and have lasting long-term effects. Indeed, Anderson, Lawton, Rexeisen and Hubbard (2006) found improvements in cultural sensitivity in just four weeks, Gilin and Young (2009) found a deeper respect for other cultures in two weeks, and Poole and Davis (2006) found improvement in personal growth in two weeks. Research would therefore suggest that overseas travel as part of global education regardless of length of stay instils a sense of ownership and empowerment, and a belief that an individual can make a difference to the world. However, a review of the literature on global citizenship in educational institutions has revealed a research gap into the experiences related to those study tours as reported by students (Hermann et al., 2016).

This brief review of the literature demonstrates the benefits associated with educational tourism and how they pertain to the development of adolescent learning. This chapter is therefore interested in understanding the socio-cultural experiences and meanings gained by adolescent students from New Zealand on an overseas educational trip.

3. Methods

The aim of this explorative study was to understand the meanings gained from the socio-cultural experiences of secondary-school students from New Zealand participating in an international school trip to Thailand. The choice of a qualitative methodology is underpinned by the philosophical perspective of interpretivism with the goal of understanding the complex world of lived experience from the point of view of those who live it (Denzin & Lincoln, 2000). The emphasis of the interviews was on an exploration of the more personal context of educational travelling, which lends itself to qualitative research (Phillimore & Goodson, 2004). Research with adolescents requires recognising and reflecting on the inequalities of power and authority between researcher and adolescent participants. Qualitative research, with its focus on the shared construction of meaning with participants and flexibility in design and process, can partially alleviate these imbalances and give participants their own voice (Bassett et al., 2008). The research was designed to allow the students to tell their stories in a safe and encouraging environment.

The explorative research was based on semi-structured interviews performed in May 2013, four months after the trip. Much can be gained from talking to teens about their personal experiences, and recruitment for this is the most challenging aspect of research with adolescents (Moolchan & Mermelstein, 2002). The students were invited to participate, with the cooperation of the school and the consent of their parents; voluntary participation was ensured. Of the 11 students who participated in the trip, nine volunteered to be part of the research (five males and four females) with two students opting not to take part for personal reasons. The 30-minute conversational interviews were conducted in a public space at the school library. This location gave the teens enough privacy to speak freely, yet with sufficient public visibility to address any concerns about safety, as recommended by Bassett et al. (2008). The main researcher is a parent herself and had a daughter participating in the school trip, so was already known to the participants. It has been suggested by Bassett et al. (2008) that prior contact with the teenagers is useful because it helps them to see the interviewer/ researcher as an approachable person. A NZ$20 gift voucher was offered to each participant to compensate them for their time and effort. Adolescents want to be taken seriously and ideally this should involve a form of incentive (Stafford, Laybourn, Hill, & Walker, 2003) rather than taking their time for granted.

A casual introduction and informal approach was used for the conversational interviews. The students were asked questions on how they became interested in the trip; what they liked and disliked about the trip, highlights and any culture shock experienced; things they found different from home; the teaching English experience; about the social aspect of travelling in a group; any personal changes they noticed since the trip; and what they would tell other students taking part in this trip next year. However, mostly the students were invited to give voice to their school-led Thailand travel experience. It was found that the participating students were open and willing to share with the researcher their unique experiences on the school trip and relished the opportunity to relate their stories and personal insights.

The interviews were digitally recorded, transcribed and manually collated and coded by the main researcher. A grounded theory approach was used for this study to discover new perspectives regarding school students' experiences and meanings gained on an international school trip. Grounded theory specifically lends itself to the discovery of relatively unknown social phenomena (Glaser & Strauss, 1967). A more flexible constructivist approach to grounded theory was used for the analysis (Charmaz, 2000) allowing for a more holistic understanding of the meanings gained by the adolescents. This was carried out through manual coding in that data were initially coded by reading through the transcripts several times while making notes that were then sorted into themes. Categories and themes developed were thus grounded in the interview data but were also compared back to the existing literature, resulting in the identification of four major themes:

1. cultural experiences;
2. teaching English experiences;
3. potential transformational experiences; and
4. educating for global citizenship.

The research is significant in that the findings can better inform secondary schools and parents about the value of school-led tourism, enhance academic tourism knowledge, and thus aid in the provision of educational and potentially transformational tourism experiences through empowering young people.

4. Findings

Apart from the more immediate cultural experiences and teaching English experiences, there were also more potentially transformational and mind-broadening experiences reported.

Cultural experiences

Of the cultural activities the students engaged in while in Chiang Mai, some of their favourites were elephant camp, Muay Thai (Thai boxing), Thai cooking school and the night markets. These cultural experiences were far reaching and centred on cultural adaptation to differences as well as acknowledgment of similarities, which allowed for cultural connection to occur:

> *It was more of an interesting culture shock like an experience in that culture.* (17-year-old male)
> *I have a broadened perspective about Asian culture.* (17-year-old female)
> *To connect with the teenage culture there that was really cool because they would take us to a mall to show us around and when it came down to it we were so alike it was unbelievable.* (17-year-old male)

The immersion into the spiritual culture of Buddhism, which is part of Thai life, led to wider questions about self and own beliefs, as illustrated below:

> *The Thai temples were incredibly spiritual ... being there was completely different, and it helped me learn a lot more about Buddhism and my religious feelings as well.* (18-year-old male)
> *Life is their religion and it made me think about like what I can do to make myself a better person in a way.* (17-year-old female)

These quotes support the notion that the cultural and spiritual experiences in Thailand took the students outside their normal realm of experiences and introduced new perspectives to their way of thinking. It broadened the students' horizons in terms of experiences available in life and cultural traditions and spiritual ways of living.

Teaching English experiences

For the New Zealand students, teaching English to underprivileged Thai students had some profound effects in terms of reflecting on their own education experiences, the rewarding experience of teaching, and gaining an appreciation for their own Western education:

> *Elaborating on the orphanages, that completely changed my way of thinking you know, how they were pleased with nothing. They were happy kids, they had gone through a lot more ... It just made me think a lot, thinking "look what I have" reflecting on myself.* (16-year-old male)
>
> *The people that I met over there like some of the small children and seeing the smiles on their faces that was really worthwhile for me. I really saw myself wanting to do something that would give something back to the community, like give back to people.* (18-year-old male)
>
> *I thought that overall it was an amazing journey and my favourite part was the English camps because the kids were great and even though I was the one teaching, I learnt from them as well. At Lamphun [orphanage] especially, they all had an intense eagerness to learn and do so much as they could in that short amount of time and I admired their desire for knowledge because in their situation, knowledge is power. It gave me a new perspective on my education that I shouldn't take my opportunities for education for granted.* (17-year-old female)

Seeing the economic hardship of life in a developing country showed these New Zealand students how valuable having an education was to the Thai people, and the commitment of the Thai students to learning became understandable. Education is perceived in Thailand as a way out of financial insecurity towards a better quality of life. For many of the New Zealand students, realisation of the difference education can make in people's lives came as a surprise, which led to potentially transformative experiences.

Potential transformational experiences

The immersion into Thai culture and education realities led to more potential long-term implications for the New Zealand students with relevance for their later study options, travel opportunities and jobs pursued, as illustrated below:

> *I worked a lot harder at school this year than I have any other year. So kind of makes me not take education for granted to know that there are kids in other countries that would kill for a chance to have an education like mine.* (17-year-old female)
>
> *I like poorer countries, I like the people, I like interacting with them, I want to bring about change as well as I want to help others. What it gave me is love for poorer countries. ... Now I would actually like to go back to one of those poorer countries, especially*

> Thailand or something like that. So it did change my view on where I would want to go in the future. (16-year-old male)
> Before this I thought: "I am so ok to stay in New Zealand for the rest of my life, I love it here" ... and now I really do have a passion to go back through Asia, especially Southeast Asia. (18-year-old male)
> My dream is to be a missionary and dentist and so maybe I would like to do dental treatment while being a missionary and go to places like that. (16-year-old female)

Some of the students experienced negative emotions and culture shock in having to adapt to a different culture but this also hinted at more transformational outcomes, as related by one male student who became ill during his time in Thailand:

> My Thai culture immersion experience was an emotional roller-coaster. I found the culture shocking, the language barrier, the homesickness and the physical sickness pretty hard to overcome but the programme itself was amazing. All the sickness and culture shock were worth it in the end as that is all part of the journey. I have had the time of my life riding elephants, learning Thai cooking, teaching English, visiting temples, etc. I don't regret any moment of it. (17-year-old male)

These more transformative experiences gained on the school trip in Thailand as a result of having gone through negative emotions and overcoming personal challenges could afterwards effect longer-term changes to the students' lives than might otherwise have occurred.

Broadening horizons and potential life-changing experiences: Educating for global citizenship

The different experiences that the New Zealand students gained not only broadened their perspectives on life as they knew it but also challenged them to reflect on their own personal goals and way of living. This led to some profound and potentially life-changing insights, which were especially important for being gained at a young age and on the cusp of adulthood:

> It's kind of something that you can't explain like how it impacts you it's just something you have to experience yourself and it kind of changes your perspective on life and makes you appreciate life a lot more than you would have originally appreciated. (17-year-old female)
> It's given me a real appreciation of what I have as well like it's my birthday in a week and mum has been saying: "what do you want for your birthday" and the only thing that I can think of is maybe a smart shirt cause I feel I have so much. Like I know I'm not rich, I come from a middle-class family but before I'd be like: "oh yeah laptop, car". And I don't have either of those but I am okay with that considering how many families don't own a car over there. Yeah I feel like it's made me a more confident person as well

'cause I was in the group with people like [names of students] who are all super confident and super intelligent and that really helped me and it also helped my leadership as well which is good. (17-year-old male)

It is a completely new experience. You went over there with the idea that you would change others but you got changed even more. I think I changed more than I changed others. It's a great experience. You are going to see a new culture, experience something you have never experienced before and you are going to hold on to that for the rest of your life. (16-year-old male)

You will see the world in a whole new way because these are people living but they are not living our way and it's just interesting to realise that there is another way to life out there. And you begin to really question if the Western way is better or that sort of thing. (17-year-old male)

The broadening and life-changing experiences gained in the Thai environment thus led to deeper and more critical reflections on Western ways of living and consuming, and a new-found appreciation of life and educational opportunities in New Zealand as part of these students' personal growth in global citizenship.

5. Discussion

This chapter explores the value of international educational school-led tourism, the meanings gained by nine New Zealand students, and any potential longer-term benefits. There has been some debate on the positive outcomes of volunteering in terms of the personal development of volunteers and improved cross-cultural understanding between hosts and guests (Guttentag, 2009) but not regarding school students. Findings from the research indicate that teaching English to Thai students and getting immersed in the Thai culture was beneficial and potentially life-changing for the New Zealand adolescents. Like Cavanagh (2012), the students attitude towards educational opportunities became more positive and consequently they worked harder along with gaining new insights into their own culture that changed their worldviews. There were then benefits in terms of their appreciation of education, respecting and relating to Thai culture, relating to Thai teenagers and developing a broadened perspective on Asian culture and Buddhist spirituality. There were also more potentially transformational and mind-broadening experiences reported, which fulfil the overall aim for this school trip, as set out by the New Zealand education agenda, of educating students for global citizenship.

Adolescents are a unique age group as, for most, an international school trip represents their first independent overseas travel experience away from their families. The new cultural and educational experiences gained are part of the process of forming mind-sets (Cullingford, 1995), which then become a background to the judgements formed about Thai people and their culture. Their overseas experiences were also potentially transformational in that the New Zealand students had to overcome personal challenges, culture confusion or intercultural adaptation (Hottola, 2004) and culture shock, and to reassess their belief systems. This highlights that the role of negative emotions needs to be considered more, despite the students reporting ultimately positive and memorable experiences. The New Zealand students thus gained subjective cultural experiences (Cushner, 2004), which are the more invisible aspects of values and habits, forming the basis for deep and meaningful learning. This led to more transformational experiences prompting changes in life philosophies, behaviours and personal beliefs (L. Brown, 2009). It is considered that more enduring transformations are a process and follow on from one's personal development and self-reflection (Kirillova, Lehto, & Cai, 2017), but this requires follow-up research.

The cultural immersion with people and culture in Thailand provided the New Zealand students with the experience of a simpler life, which led to them questioning consumerism – the preoccupation with materialism and owning things – in the West. Like Baudrillard (1998), some of the students questioned the consumer society of the West where the consumption of objects has become the new morality replacing spiritual endeavours or more basic human rights motivations. This reflects a deeper level of engagement for these students from New Zealand and goes against some of the criticism levelled at supposedly rich, over-indulged students from the West taking school trips to poorer countries.

6. Conclusions

Not much is known about the first immersions of young people into a different culture and foreign education system, and the personal changes that these experiences can engender. Overall, educational school-led tourism is a poorly researched and little understood segment within tourism studies, not only regarding its scale but also its specific nature and needs (Campbell-Price, 2014; Dale et al., 2012), and what it means for the students involved. Adolescents are perceived as a tough group to engage in research, but ensuring that their voices are being heard allows for deeper insights on social issues in tourism and educational initiatives at schools in distinctive ways. What transpired from the in-depth interviews was that immersion into Thai culture and extensive contact with Thai people over a three-week period led to some profound life-changing and mind-broadening experiences for the New Zealand students who participated in the school trip. The subjective cross-cultural learning experiences were particularly deep because of the high level of interaction with Thai students and Thai culture (van 't Klooster et al., 2008), which involved elements of cultural adaptation and culture shock comprising negative emotions. Students reported personal development and increased social responsibility (Wakeford, 2013); improved confidence and transformed relationships within the travel party (Saitow 2009); and more developed social behaviour and maturity (Peacock 2006). But mostly, the school trip experiences in Thailand induced personal and critical reflections about the meaning of life and the value of educational opportunities, as well as being a catalyst for spiritual change in at least some of the students. The outcome, then, was an increased valuing of education but also a partial rejection of consumer society because of a more global awareness of the opportunities and pitfalls of life in Western societies compared with developing and Asian countries. This can prove that there are numerous social benefits to educational tourism not related to the curriculum but also that there are negative experiences and unintended outcomes.

A deeper connection between cultures was achieved leading to increases in global citizenship as identified through changes in consumer behaviour and life aspirations or, in other words, a more spiritual transformation. This is evidence of educational travel achieving an increased understanding of global citizenship which differs from findings by Hermann et al. (2016) who reported little reflection on cultural immersion by tertiary students. However, the current study is at a secondary school level and features a trip that focused on global ethical issues and cultural connection, which perhaps is better termed as 'sprirual citizenship'. Global spiritual citizenship proposes a positive outcome to the 'globalisation of the people' by emphasising a realisation of common human values and social connections. This kind of school-led tourism, then, can promote global citizenship and optimism for a better world through firsthand experiences of teaching disadvantaged students in another country and being immersed in their culture. In a rapidly changing and interdependent world, school-led tourism can make a difference by engaging students in life-shaping experiences bringing personal moral development, but more follow-up research is needed to assess the long-term implications of overseas educational trips for enduring transformational experiences. There appears to be social and ethical value in the significant expense and time commitment of school-led tourism, which benefits the adolescents, parents, teachers and the overall education agenda of educating students for global citizenship, thereby promoting a more tolerant and just society.

References

Anderson, P.H., Lawton, L., Rexeisen, R.J., & Hubbard, A.C. (2006). Short-term study abroad and intercultural sensitivity: A pilot study. *International Journal of Intercultural Relations, 30*(4), 457–469.

Ballantyne, R., & Packer, J. (2002). Nature-based excursions: School students' perceptions of learning in natural environments. *International Research in Geographical and Environmental Education, 11*(3), 218–236.

Bassett, R., Beagan, B., Ristovski-Slijepcevic, S., & Chapman, G. (2008). Tough teens: The methodological challenges of interviewing teenagers as research participants. *Journal of Adolescent Research, 23*(2), 119–131.

Baudrillard, J. (1998). *The consumer society: Myths & structures*. London, England: Sage.

Beard, C.M., & Wilson, J.P. (2006). *Experiential learning: A best practice handbook for educators and trainers* (2nd ed.). London, England. Kogan Page Publishers.

Bodger, D. (1998). Leisure, learning, and travel. *Journal of Physical Education, Recreation & Dance, 69*(4), 28–31.

Brown, L. (2009). The transformative power of the international sojourn: An ethnographic study of the international student experience. *Annals of Tourism Research, 36*(3), 502–521.

Brown, S. (2005). Travelling with a purpose: Understanding the motives and benefits of volunteer vacationers. *Current Issues in Tourism, 8*(6), 479–496.

Byrnes, D.A. (2001). Travel schooling: Helping children learn through travel. *Childhood Education, 77*(6), 345–350.

Campbell-Price, M. (2014). *International school trips: A critical analysis of multiple stakeholder perspectives* (Unpublished doctoral thesis). Otago University, Dunedin, New Zealand.

Carr, N., & Cooper, C. (2003). Schools' educational tourism. In B.W. Ritchie (Ed.), *Managing educational tourism* (pp. 130–180). New York, NY. Channel View Publications.

Cavanagh, C. (2012). *To what extent has the impact of overseas studying changed education student's perceptions on higher education?* Retrieved May 20, 2015, from https://www.hope.ac.uk/media/liverpoolhope/contentassets/documents/media,15635,en.pdf

Charmaz, K. (2000). Grounded theory: Objectivist and constructivist methods. In N.K. Denzin & Y.S. Lincoln (Eds.), *Handbook of qualitative research* (2nd ed., pp. 509–535). Thousand Oaks, CA: Sage.

Chen, L.-J., & Chen, J.S. (2011). The motivations and expectations of international volunteer tourists: A case study of 'Chinese village traditions'. *Tourism Management, 32*(2), 435–442.

Cohen, E.H. (2014). Self-assessing the benefits of educational tours. *Journal of Travel Research, 55*(3), 353–361.

Coleman, J.A. (1997). Residence abroad within language study. *Language Teaching, 30*(1), 1–20.

Cooper, C., & Latham, J. (1988). English educational tourism. *Tourism Management, 9*(4), 331–334.

Cullingford, C. (1995). Children's attitudes to holidays overseas. *Tourism Management, 16*(2), 121–127.

Curzon, L.B. (1997). *Teaching in further education: An outline of principles and practice* (5th ed.). London, England: Cassell.

Cushner, K. (2004). *Beyond tourism: A practical guide to meaningful educational travel.* Oxford, England. Scarecrow Education.

Dale, N., Ritchie, B., & Keating, B. (2012). Understanding constraints and their impact on school excursion tourism. *Tourism Analysis, 17*(6), 805–812.

Denzin, N.K., & Lincoln, Y.S. (Eds.). (2000). *Handbook of qualitative research* (2nd ed.). Thousand Oaks, CA. Sage.

Dwyer, M. (2004). More is better: The impact of study abroad program duration. *Frontiers: The Interdisciplinary Journal of Study Abroad, 10*, 151–163.

Eckert, J., Luqmani, M., Newell, S., Quraeshi, Z., & Wagner, B. (2013). *Developing short-term study abroad programs: Achieving successful international student experiences.* Retrieved July 12, 2016, from http://search.proquest.com/openview/c449710e4d01577119b954ef6f45b1c8/1?pq-origsite=gscholar

Falk, J., & Dierking, L. (1997). School field trips: Assessing their long-term impact. *Curator: The Museum Journal, 40*(3), 211–218.

Ferrante, J. (2013). *Sociology: A global perspective* (8th ed.). Belmont, CA: Wadsworth Cengage Learning.

Fischer, K. (2009). *Short study abroad trips can have lasting effect, research suggests.* Retrieved June 22, 2015, from http://web3.coehs.siu.edu/SSW/ecuador/Short_Study_Abroad_Trips_Can_Have_Lasting_Effect_Research_Suggests.pdf

Forsey, M., Broomhall, S., & Davis, J. (2012). Broadening the mind? Australian student reflections on the experience of overseas study. *Journal of Studies in International Education, 16*(2), 128–139.

Fox, G. (2011, May 31). Are these school trips just pricey packages? *The Telegraph*. Retrieved May 20, 2015, from http://www.telegraph.co.uk/finance/property/schools/8541832/Are-these-overseas-school-trips-just-pricey-package-holidays.html

Frändberg, L. (2010). Activities and activity patterns involving travel abroad while growing up: The case of young Swedes. *Tourism Geographies: An International Journal of Tourism Space, Place and Environment, 12*(1), 100–117.

Gilbertson, K. (2006). *Outdoor education: Methods and strategies*. Champaign, IL: Human Kinetics.

Gilin, B., & Young, T. (2009). Educational benefits of international experiential learning in an MSW program. *International Social Work, 52*(1), 36–47.

Glaser, B.G., & Strauss, A.L. (1967). *The discovery of grounded theory: Strategies for qualitative research*. New York, NY: Aldine.

Greene, S., & Hogan, D. (2005). *Researching children's experience: Approaches and methods*. London, England: Sage.

Griffin, J. (1998). Learning science through practical experiences in museums. *International Journal of Science Education, 20*(6), 655–663.

Grover, S. (2004). Why don't they listen to us? On giving power and voice to children participating in social research. *Childhood: A Global Journal of Child Research, 11*(1), 81–93.

Guttentag, D.A. (2009). The possible negative impacts of volunteer tourism. *International Journal of Tourism Research, 11*(6), 537–551.

Hermann, I., Meijer, K., & Van Koesveld, S. (2016). Global citizenship, tourism and consumerism: A narrative enquiry into the global consumer-citizen spectrum in students' study tour experiences. *Hospitality & Society, 6*(2), 131–151.

Hottola, P. (2004). Culture confusion: Intercultural adaptation in tourism. *Annals of Tourism Research, 31*(2), 447–466.

Houser, C., Brannstrom, C., Quiring, S.M., & Lemmons, K.K. (2011). Study abroad field trip improves test performance through engagement and new social networks. *Journal of Geography in Higher Education, 35*(4), 513–528.

Ibbotson, L. (2014, June 3). Ambitious travel plans take shape. *Otago Daily Times*. Retrieved from https://www.odt.co.nz/regions/queenstown-lakes/ambitious-travel-plan-takes-shape

Kennedy, H. (1994). *Learning works: Widening participation in further education*. Coventry, England: The Further Education Funding Council. Retrieved from http://dera.ioe.ac.uk/15073/2/Learning%20works%20-%20widening%20participation%20in%20further%20education%20%28Kennedy%20report%29.pdf

Kirillova, K., Lehto, X., & Cai, L. (2017). What triggers transformative tourism experiences? *Tourism Recreation Research, 42*(4), 498–511.

Kolb, D. (1984). *Experiential learning: Experience as the source of learning and development.* Englewood Cliffs, NJ: Prentice Hall.

Landis, D., & Brislin, R.W. (2013). *Handbook of intercultural training: Area studies in intercultural training.* New York, NY: Elsevier.

Larsen, S., & Jenssen, D. (2004). The school trip: Travelling with, not to or from. *Scandinavian Journal of Hospitality & Tourism, 4*(1), 43–57.

McMillian, R., & Opem, G. (2002). *Study abroad: A lifetime of benefits.* Retrieved June 23, 2015, from http://www.iesabroad.org/study-abroad/news/study-abroad-lifetime-benefits

Ministry of Education. (2007a). *The international education agenda 2007–2012.* Retrieved from http://www.minedu.govt.nz/~/media/MinEdu/Files/EducationSectors/InternationalEducation/FormsAndGuidelines/11950%20SummaryOfInternationalEducationAgenda%20100807.pdf

Ministry of Education. (2007b). *The New Zealand curriculum.* Wellington, New Zealand: Author. Retrieved from http://www.minedu.govt.nz/Boards/TeachingAndLearning/NewZealandCurriculum.aspx

Ministry of Education. (2014). *Education outside the classroom.* Retrieved March 4, 2015, from http://eotc.tki.org.nz/

Moolchan, E.T., & Mermelstein, R. (2002). Research on tobacco use among teenagers: Ethical challenges. *Journal of Adolescent Health, 30*(6), 409–417.

Nunan, P. (2005). *An exploration of the long term effects of student exchange experiences.* Retrieved May 25, 2015, from http://aiec.idp.com/uploads/pdf/Nunan%20%28Paper%29%20Thurs%200900%20MR5.pdf

O'Callaghan, J. (2006). *How one semester study abroad experience affects an undergraduate college student's identity development* (Unpublished master's thesis). Montana State University, Bozeman, MT.

Paige, M., Fry, G., Stallman, E., Jon, J.-E., & Josic, J. (2009). *Beyond immediate impact: Study abroad for global engagement,* Intercultural Education, 55, 29–44 doi: 10.1080/14675980903370847

Peacock, A. (2006). *Changing minds: The lasting impact of school trips.* Exeter, England. University of Exeter Press.

Perry, L., Stoner, L., & Tarrant, M. (2012). More than a vacation: Short-term study abroad as a critically reflective, transformative learning experience. *Creative Education, 3*(5), 679–683.

Petty, G. (1998). *Teaching today.* Cheltenham, England: Redwood Books.

Phillimore, J., & Goodson, L. (Eds.). (2004). *Qualitative research in tourism: Ontologies, epistemologies and methodologies*. London, England: Routledge.

Poole, D.L., & Davis, T.S. (2006). Concept mapping to measure outcomes in study abroad programs. *Social Work Education, 25*(1), 61–77.

Price, S., & Hein, G. (1991). More than a field trip: Science programmes for elementary school groups at museums. *International Journal of Science Education, 13*(5), 505–519.

Richter, L.K. (2005). Not a minor problem: Developing international travel policy for the welfare of children. *Tourism Analysis, 10*(1), 27–36.

Ritchie, B.W. (2009). School excursion management in national capital cities. In R. Maitland & B. Ritchie (Eds.), *City tourism: National capital perspectives* (pp. 185–200). Wallingford, England: CABI.

Rivers, J. (2006). *Effectiveness of programmes for curriculum-based learning experiences outside the classroom: A summary*. Wellington, New Zealand: Ministry of Education.

Rivers, W. (2008). Is being there enough? The effects of homestay placements on language gain during study abroad. *Foreign Language Annals, 31*(4), 492–500.

Saitow, A. (2009). *Educational travel and adolescent learning: A theory* (Unpublished PhD thesis). University of Montana, Missoula, MT.

Schänzel, H.A., & Smith, K.A. (2014). The socialization of families away from home: Group dynamics and family functioning on holiday. *Leisure Sciences, 36*(2), 1–18.

Scoffham, S., & Barnes, J. (2011). Happiness matters: Towards a pedagogy of happiness and well-being. *Curriculum Journal, 22*(4), 535–548.

Small, J. (2008). The absence of childhood in tourism studies. *Annals of Tourism Research, 35*(3), 772–789.

Stafford, A., Laybourn, A., Hill, M., & Walker, M. (2003). "Having a say": Children and young people talk about consultation. *Children & Society, 17*(5), 361–373.

Taras, V., & Gonzalez-Perez, M.A. (2014). *The Palgrave handbook of experiential learning in international business*. London, England: Palgrave Macmillan.

Tarrant, M.A., Lyons, K., Stoner, L., Kyle, G.T., Wearing, S., & Poudyal, N. (2014). Global citizenry, educational travel and sustainable tourism: Evidence from Australia and New Zealand. *Journal of Sustainable Tourism, 22*(3), 403–420.

van 't Klooster, E., van Wijk, J., Go, F., & van Rekom, J. (2008). Educational travel: The overseas internship. *Annals of Tourism Research, 35*(3), 690–711.

Wakeford, S. (2013). *The motivations, expectations and experiences of secondary school students involved in volunteer tourism: A case study of the Rangitoto College Cambodia house building trip* (Unpublished PhD thesis). Auckland University of Technology, New Zealand.

Wearing, S. (2001). *Volunteer tourism: Experiences that make a difference.* Wallingford, England: CABI.

Williams, T.R. (2005). Exploring the impact of study abroad on students' intercultural communication skills: Adaptability and sensitivity. *Journal of Studies in International Education, 9*(4), 356–371.

Xplore Camps. (2017). *Education or vacation?* Retrieved August 29, 2017, from http://www.xploreactivitycamps.com/education-or-vacation/

Elricke Botha, Petrus van der Merwe and Melville Saayman

Interpretation in Protected Areas: The case of the Kgalagadi Transfrontier Park, South Africa

1. Introduction

Protected areas play a fundamental role in biodiversity conservation (Woodley et al., 2012). A protected area is an "area of land and/or sea especially dedicated to the protection and maintenance of biological diversity, and of natural and associated cultural resources, and managed through legal or other effective means" (Dudley & Stolton, 2008:9). Protected areas such as wilderness areas, conserved community areas, nature reserves and national parks (IUCN, 2018) also have a strong visitor/tourist component. Weaver and Lawton (2017) explain that parks, as protected areas, were first developed *for* tourists' enjoyment of scenery and, in some cases, for hunting, with both activities having minimal consideration for ecological impacts. Later on, the development of parks *with* tourists emerged and this is still ongoing, as managers realised the considerable impact humans can have on the environment and have therefore opted for more support for conservation (Weaver & Lawton, 2017). These authors, however, recognised a third-generation situation where parks *and* tourists coexist in a relationship in which both parties obtain benefits. On the one hand, tourists fund the operational costs of conserving the environment due to a decrease in public funding worldwide (Eagles, 2002; 2014) and on the other hand, protected areas provide ample opportunity for nature tourists to satisfy their specific needs.

As numerous authors have found, nature tourists to protected areas (referred to hereafter as ecotourists) tend to be well educated (Bidder, Kibat & Fatt, 2016; Botha, Saayman & Kruger, 2016; Cheung & Jim, 2013; Cheung & Jim, 2014; Jurdana, 2009; Kruger & Saayman, 2010) and have relatively high levels of disposable income (Cheung & Jim, 2014; Sharpley, 2006). They also place a high value on biodiversity preservation (Cheung & Jim, 2013; Sheena et al., 2015), enjoy nature (Cheung & Jim, 2013; Sheena, Mariapan & Aziz, 2015) and are concerned with environmental issues (Ballantyne, Packer & Falk, 2011; Sheena et al., 2015). Although ecotourists may already have a high commitment to the environment and engage in sustainable environmental practices (Ballantyne et al., 2011), they tend to have high expectations for quality information or learning opportunities (Bidder et al., 2016; Chen & Jim, 2012; Cheung & Jim, 2013; Sheena et al., 2015). It is for this reason that managers of protected areas usually make use of interpretation services to not only satisfy tourists' information needs, but also use it as a means to conserve the environment through education. As will be explained later, interpretation is an educational activity that facilitates an enjoyable experience (Moscardo, 1999) that is typically delivered through various media. As stated by Baba Dioum (1968), "in the end we will conserve only what we love; we will love only what we understand; and we will understand only what we are taught".

Although ecotourists typically portray the usual traits (for example an appreciation for nature or travelling to relax and to escape daily routine to name but a few) it can differ from one context to the other (Sheena et al., 2015). It is precisely for this reason that interpretation planning also includes the identification of visitors (i.e., ecotourists) (Ham, Housego & Weiler, 2005). Other planning steps include interpretive inventory, identification of interpretive goals, determination of the outcomes of the goals, development of the themes, development of media matrices, implementation and evaluation (Ham et al., 2005). Understanding ecotourists' specific needs, characteristics, motivations and behaviour, especially within the context of interpretation, assists with tailor-making nature products and services that satisfy these specific needs and, in return, sustains funding for conservation (Lück, 2015).

The aim of this chapter is therefore to give an analysis of interpretation and, through a case study, provide practical guidelines on how interpretation can be incorporated practically in protected areas.

2. Literature Review

The following review focuses on the key literature about interpretation in protected areas.

2.1 *Interpretation*

From an analysis of the available definitions of interpretation, one can conclude that interpretation is the communication of the significance of the place people are visiting (Moscardo, 1999), educating them on the meanings and relationships in an enjoyable manner (Orams, 1996) that creates a rewarding visitor experience and, within an environmental context, leads to positive behavioural change (Moscardo & Ballantyne, 2008). This is achieved through showing them original objects and illustrative media, as well as providing first-hand experiences (Tilden, 1977; Ward & Wilkinson, 2006), rather than merely communicating factual and technical information (Ham, 1992; Ward & Wilkinson, 2006). Although interpretation is used in numerous contexts such as museums, wine-tastings or art galleries, it has a specific role within environmental contexts. It is clear from the above definition that interpretation within an environmental context goes beyond the mere communication of information that one would expect in the other contexts. Education, revealing the *meanings* and *relevance* of specific issues, is required for potential positive behavioural change, related to that issue, to take effect.

Jacobs and Harms (2014) indicated a strong correlation between interpretation and behavioural change or intentions, whereas other studies (Ballantyne et al., 2011; Ballantyne, Packer & Hughes, 2009; Powell & Ham, 2008) queried this causal relationship. Understanding the causes of behaviours and how they are formed are therefore fundamental to the design of interpretation.

2.2 Principles for interpretation design

As will become clear in this section, many researchers have made use of different concepts and theories to explain how interpretation works and therefore how to develop it (Botha et al., 2016). It is thus crucial to explain these concepts and theories first before focussing on interpretation and behaviours. It is, though, not the aim of this chapter to provide a full explanation of all the related theories, but rather to provide a brief explanation of each.

Due to the popularity of the use of interpretation in environmental contexts and thus the emphasis on conservation, many authors (Ballantyne & Hughes, 2006; Ballantyne, Packer, Hughes & Dierking, 2007; Hughes, Ham & Brown, 2009; Moscardo, Woods & Saltzer, 2004) have researched interpretation from the perspective of a persuasive communication lens. By referring to the communication process (i.e. sender to receiver), persuasive communication is the process during which the receiver's attitude or behaviour is voluntarily formed, changed or sustained by what the sender intended with the message (De Wet, 2010). Since pure persuasion does not require the persons to understand the reason for their actions (De Wet, 2010), interpretation should rather focus on education, a form of persuasion that helps the receiver understand.

Literature on interpretation and education has mainly focussed on the different educational theories such as Pavlov's (1958) classical conditioning theory, in which learning occurs through association, or Thorndike (1911) and Skinner's (1935) operant conditioning, where learning situations are created in which the stimuli are direct consequences of one's behaviour. Little empirical results are available on interpretation and these conditioning theories, but Orams (1994) and Kuo (2002) propose that these types of theories are typically employed by managers of protected areas as hard interpretation. Such interpretation might be physical (e.g., animal viewing platforms or pathways), regulatory (e.g., punishment or reward for certain behaviours) and economic (e.g., using fees as an incentive) (Kuo, 2002).

On the other hand, Bandura's (1977) theory, learning through modelling, proposes that learning can occur by observing other people's behaviour through paying attention, recalling the observation, and then being able to reproduce the observation and motivated to imitate the behaviour. This educational theory has not been researched much in the interpretation context either, but Wiener, Needham and Wilkinson (2009) explain that tourism businesses should "practise what they preach" to refrain from confusing the guests and discarding the conservation messages. It is a typical example of how the observational theory applies to protected areas.

The best-known educational theory, typically employed in interpretation, is Piaget's (1972) cognitive learning theory. This theory proposes that the cognitive system naturally seeks equilibrium through assimilation (i.e., new information deduced to being consistent with existing schemas) and accommodation (i.e., adaption of existing cognitive schemas with new information). Tourism businesses typically provide educational information as interpretation to guests, educating them on specific matters of importance to the destination. However, it is exactly here where interpretation within the environmental context aims to be different. Merely educating or providing information will not suffice, as interpretation in the environmental context has a specific goal attached to it, namely pro-conservational behaviour.

When providing educational information with a specific pro-conservational goal attached to it, managers of protected areas should keep Festinger's (1957) cognitive dissonance theory in mind. The cognitive dissonance theory states that people want to maintain a state of consonance, but when in a state of dissonance, people will either change behaviour or rationalise their refusal to change behaviour (Eunson, 2008). Educational messages should therefore rather opt to change behaviour to pro-conservation than to create a situation where the tourists will opt not to change their behaviour in line with conservation practices. When one refers to pro-conservational behaviour, managers of protected areas should bear in mind that several behaviours can be regarded as environmentally significant.

Stern (2000) explains that environmentally significant behaviour includes environmental activism (e.g., involvement in environmental organisations), non-active behaviour in the public sphere (e.g., indirectly through acceptance of environmental policies), private-sphere environmentalism (e.g., purchasing environmentally friendly products as well as responsible use of products such as heaters, water and recycling), and other behaviour (e.g., building with 'greener' products). It is therefore necessary to link interpretation to the specific behaviour the protected area managers would like to achieve. Merely presenting information to guests will not be sufficient if such information does not direct the guest to a certain type of behaviour.

Stern (2000) also asserts that single-variable explanations (e.g., only attitudes) for informed behavioural change are limited, since behaviour is determined by multiple variables. Ajzen's (1991) theory of planned behaviour (TPB) is a typical example of what Stern means. The TPB (Ajzen, 1991) indicates that (i) the attitude towards a behaviour (favourable or unfavourable), (ii) the subjective norms (perceived social pressures), and (iii) the perceived behaviour control (ease or difficulty to perform the behaviour) all predict behavioural intentions that lead to actual behaviour. As previously indicated, ecotourists are already inclined to place a high value on preservation and have concern about environmental issues; it can thus be assumed that their attitudes towards pro-conservation behaviour are generally positive.

Considering that ecotourists also tend to visit natural areas like protected areas, these destinations can influence guests (i.e., subjective norms) to engage in pro-conservation behaviour by staff performing the behaviour as well (e.g., staff engage in recycling). Interestingly, though, the theory explains that perceived behavioural control and behavioural intention directly predict behaviour (Ajzen, 1991). If manager of protected areas therefore develop interpretation that overcomes the barriers to perform the behaviour (i.e., high-perceived control), the guests will be able to perform the behaviour. Managers of protected areas should thus be more precise in their communication about the expected behaviour. This is also highlighted by Ham et al.'s (2005) interpretation planning phases in which they indicated that the outcomes of the interpretive goals should be determined.

Most of the interpretation services currently offered in protected areas aim to enhance the tourists' experience (Ballantyne et al., 2011) or create awareness of a specific conservational issue. The idea is not to discard the importance of these aims, but rather to take them a step further. If protected areas are mandated to protect and maintain biological diversity, there should at least be interpretation that includes methods to create such behaviour. For example, if the protected area is protecting a specific endangered animal, the interpretation could include behavioural methods such as volunteering programmes or platforms to contribute financially to the programme. These types of interpretation services give life to the 'parks *with* tourists' concept, as mentioned by Weaver and Lawton (2017).

Successful interpretation contributes numerous benefits to protected areas such as increased environmental awareness and/or behaviour, but also increased satisfaction, loyalty, purchasing, and revenue (De Rojas & Camarero, 2008; Hwang et al., 2005; Lee, 2009; Zeppel & Muloin, 2008). As previously indicated, ecotourists have high expectations for quality information or learning opportunities (Lück, 2003). In an environment where protected areas are competing for the same type of tourists' contributions (i.e., financially and conservation-wise), it is necessary for these areas to provide an interpretation that also satisfies the high-quality needs of the tourists. Although quite old, Ham's (1992) EROT model is still applicable today. Ham explains that interpretation should be designed to be enjoyable (i.e., holding the attention), relevant (i.e., creating meaning and personal interpretation), organised (i.e., easy to follow with minimal effort) and themed (i.e., qualities of a story). These principles have spurred numerous research studies, for example testing short-term and long-term tourism experiences (Ballantyne et al., 2011); the threshold for behavioural requests (Smith et al., 2012); the effectiveness of signage (Marchall, Granquist & Burns, 2017); and the role of emotional displays (Wijeratne et al., 2014).

It is therefore safe to say that designing and offering high quality interpretation is not a straightforward activity. Based on the discussed theories, interpretation has different forms in which it can be employed and requires managers to be aware of new needs and developments. Managers of protected areas will have to provide amalgamated interpretation services if they want to remain relevant in a competitive environment and fulfil their mandate of conservation.

2.2.1 Interpretation examples

As previously explained, interpretation can take on numerous forms that can easily be explained by a continuum. On the one end, soft interpretation examples include services and facilities that illustrate precisely what interpretation aims to accomplish (i.e., education) and on the other end, hard interpretation includes examples that are not readily identified as interpretation but has an impact on the experience (Kuo, 2002; Orams, 1994).

Soft interpretation

Perhaps the most prominent soft interpretation example is a visitor centre that displays information on a park. Displays may include elements of personal (i.e., person to person) or impersonal (i.e., without direct contact) messages.

Personal interpretation can take the form of guided tours, lectures or information sessions on a specific topic, or answering queries that visitors may have. Most of the time, park managers make use of impersonal interpretation through various media. This may include guide books, exhibits, name plates on trees, bird hides, life-size examples of animals, podcasts and games, as well as panels, posters, videos, maps, illustrations, and appropriate cellular applications.

Hard interpretation

Hard interpretation is typically referred to as 'visitor management strategies'. As previously explained, these strategies are categorised as physical, regulatory and economic. Physical interpretation includes zoning, pathways, restricting access to specific areas by fencing them in, or limited car parks. Regulatory interpretation includes rules and regulations communicated in the guide book, restriction or warning signs and speed limitation signs. Economic interpretation includes higher entrance fees in high seasonality seasons or lower fees for pro-conservational efforts, fines for littering, or conservation fees.

2.3 Case Study

The following case study is a discussion on interpretation in some protected areas. The case study is based on the Kgalagadi Transfrontier Park (KTP), situated in South Africa. The reason for selecting this park is that it had been identified by the South African National Parks' (hereafter referred to as SANParks) management as a priority park for developing interpretation services (Moore, 2017).

2.3.1 Background

Kgalagadi Transfrontier Park is one of the 19 national parks managed by SANParks and is unique in the sense that it is a transfrontier park extending into Botswana (Figure 1). Transfrontier parks or transfrontier conservation areas (TFCA) are defined as "relatively large areas, straddling frontiers between two or more countries and covering large-scale natural systems encompassing one or more protected areas" (South Africa, 2004). The park was the first TFCA in

Southern Africa and was opened on 12 May 2000 (South Africa, 2004). Two of the most important cultures in the area are the Khomani culture (the last surviving indigenous San/Bushmen communities in South Africa), whose members used to inhabit its southern section, and the !Xam culture, whose members live south of Upington (UNESCO, 2017). For this reason, the park was proclaimed a World Heritage Site in 2017.

At present, the park has only three information kiosks, at the Twee Rivieren, Nossob, and Mata Mata rest camps. These information kiosks, however, are very basic, with nothing but a few displays and posters (Moore, 2017). It is against this backdrop that park management has considered offering interpretation services based on visitors' needs and preferences. The main aim of the research was therefore to provide answers to the following questions in order to develop relevant interpretation services for the KTP:

- What are the interpretation preferences of tourists visiting the national park?
- What are the themes or types of information visitors are interested in?
- Do visitors prefer specific media through which the themes or information should be presented?

2.3.2 Methods

To achieve these aims, the following approach was implemented: A questionnaire was developed by TREES (Tourism Research in Economic Environs and Society) at the North-West University in cooperation with the management at SANParks. Convenience sampling, a non-probability sampling method, was used to conduct the survey. Convenience sampling means that the researchers include participants who are easy to access (Maree & Pietersen, 2007). The questionnaire was developed to be distributable physically as well as online. The hard-copy questionnaire was distributed in the Kgalagadi Transfrontier Park from 15 to 19 January 2017, whereas a link to the questionnaires was distributed by the SANParks by sending a link to the questionnaire to those overnight visitors who stayed at the park in the previous 12 months. In total, 562 visitor questionnaires were completed, of which 520 were completed online and 42 were completed by guests in the park.

2.3.3. Results and discussion

The results and discussion of the case study consist of the following: First, the importance of the different types of information and services; second,

Figure 1: Kgalagadi Transfrontier Park
Source: SANParks (2018)

the importance of specific media in interpretation centres; and third, ways in which information should be presented will be discussed.

Importance of information/services for the park

Respondents' preferences were captured on a five-point Likert-type scale (ranging from 1 = *not at all important* to 5 = *extremely important*). Many services and forms of information were found to be important, but the following were viewed as the most important:

- Knowledgeable staff who can handle any queries (4.44)
- Park rules and regulations communicated through signs and brochures (4.35)
- Map of the park provided at no additional cost (4.28)
- Static displays in bird hides in the park Bird hides in the park through static displays (4.28)
- Relevant information and activities of the park, shown on the website (4.27)

Practical examples

Tourists' experiences begin well before visiting the park. Providing the necessary interpretation services or information on the park's website will help create anticipation of visiting the park. These web interpretation services can, as explained later, include visual and audio media to capture the attention of the tourists by posting short videos of the latest sightings in the park. Some parks also include game ranger podcasts about certain events that they have experienced, for example, witnessing how poachers were caught or the birth of a rhinoceros calf. The website and knowledgeable staff are aspects referred to as 'secondary interpretation' and maps, park rules and regulations (see Figure 2) are referred to as 'tertiary interpretation'.

Stewart and colleagues (1998) explain that secondary interpretation is auxiliary to a wider activity that is not readily identifiable as interpretation, but is an integral feature of the interpretation. They also explain that tertiary interpretation is not always considered to be an interpretation, as it is indistinct as an interpretive activity but may impact on the experience. These aspects are therefore valuable in creating a good quality tourist experience.

As protected areas are mandated to protect the environment, it is obvious that all staff members need to work together on all frontiers to achieve this aim. This can only be accomplished by all staff members understanding the concept of conservation and how their actions contribute towards this goal. Every staff member, from check-in to cleaning, should be able to answer the basic interpretation queries or at least make an effort to find the answer on behalf of the visitor.

Rules and regulations (Figures 2 and 3) typically have a role to play in the conservation of the animals (e.g., speed limits) or as a means to protect the visitor (e.g., warning tourists about certain behaviour that would provoke animals to attack). Parks, however, fail to inform tourists about the reasoning behind the rules and regulations; if tourists are made aware of them, they might be more supportive (Kuo, 2002) and comply with the desired behaviour. The KTP can provide this information in the form of a park guide/booklet. This guide

Table 1: Importance of information/services for the park

	ASPECTS	mean	*Level of importance
1	Game drives with interactive field guides	3,26	Neither very important nor less important
2	Map of the Park provided at no additional cost (because of safety/security/information reasons)	4,28	Very important
3	Information board with sightings of the day	3,98	Very important
4	Identification of trees (e.g., nameplates or information boards)	4,12	Very important
5	Static displays in bird hides in the park Bird hides in the park through static displays	4,28	Very important
6	Static displays in lookout point in the park Lookout points in the park through static displays	4,21	Very important
7	Knowledgeable staff who can handle any queries (general information as well as specialised information on the park)	4,44	Very important
8	Relevant information on and activities in the park, shown on the website	4,27	Very important
9	Experience of the local culture/community's culture or way of life	3,27	Neither very important nor less important
10	Cuisine experiences (e.g., game meat dishes or local culture's prepared dishes)	3,12	Neither very important nor less important
11	Event nights (e.g., star gazing sessions, educational movies on nature/cultures, photography sessions)	3,53	Very important
12	Additional books, magazines or newspapers to be bought for supplementary information on the environment, conservation or astrology	3,59	Very important
13	Clear directions in the park (e.g., to picnic areas and rest camps, and travelling time)	4,18	Very important
14	Park rules and regulations communicated through signs and brochures	4,35	Very important

*Five-point Likert-type scale ranging from 1 = *Not at all important* to 5 *Extremely important*.

Figure 2: Regulatory signage (Photos: P. van der Merwe)

Figure 3: Regulations in Kgalagadi Transfrontier Park (Photo: P. van der Merwe)

can include the most basic information, from contact details, rules and regulations (and the reasoning behind it) and animal lists to a map of the park with interesting sightings and other interpretive material. It is crucial that this guide should be designed in a manner that appeals to the eye and that information should be brief enough to keep the attention of the visitor.

Avitourists or birders are a specific niche of tourists with specialised knowledge on birds as well as resources (books) and equipment (cameras) to accompany the activity. Usually, bird hides only include posters with information on birds, but in some cases, some of these posters are not region specific. It is important that the information presented is relevant by including bird species that are located in the area. Due to the nature of the activity, bird hides can be designed for an optimal experience. Usually, these bird hides are built as wooden structures over water, which adds to the noise level when tourists arrive or leave. Noise can scare the birds away and the physical structure should thus be designed for minimal noise. Due to the number of resources and equipment these tourists carry along, it is preferable that the design of the bird hide incorporates some element of comfort in the form of camera rests and tables for books.

Importance of specific media in interpretation centres
Respondents were asked to indicate the importance of certain types of media and best practices for interpretation services. Their preferences were captured on a five-point Likert-type scale (ranging from 1 = *not at all important* to 5 = *extremely important*). The following types of media obtained the highest mean values:

- Visual media (3.86)
- Printed media (3.79)
- Easily readable text or information in relevant media (3.74)
- 2-D displays (e.g., posters with information on the park) (3.62)
- Audio media (3.61)

The following good principles/best practices obtained the highest mean values:

- The staff of the interpretation visitor centre should be able to answer any question (4.15).
- Easily readable text or information in relevant media (e.g., not too small font) (3.74)
- Display of props or objects (e.g., archaeology, plant, insect or cultural displays) (3.61)
- Concise information presented in the media (3.57)

Table 2: Importance of media and good principles/best practices

	ASPECTS	mean	*Level of importance
	TYPES OF MEDIA		
1	Visual media (e.g., pictures and diagrams that present interesting information on the park)	3,86	Very important
2	Audio/visual media (e.g., videos that explain the history of the park or evolution)	3,45	Neither very important nor less important
3	Audio media (e.g., bird or animal sounds)	3,61	Very important
4	Virtual reality (e.g., virtual tours through traditional houses of cultures)	3,07	Neither important nor less important
5	Printed media (e.g., brochures of the interpretation visitor centre and posters with interesting information)	3,79	Very important
6	Technology-based media (e.g., interactive touch screens that illustrate information you would like to know)	3,37	Neither important nor less important
7	Game-like media (e.g., short quizzes, puzzles and "did you know?"-facts)	2,98	Neither important nor less important
8	2-D displays (e.g., posters with information on the park)	3,62	Very important
9	3-D displays (e.g., life-size examples of animals or cultural artefacts, or virtual tours)	3,31	Neither important nor less important
	GOOD PRINCIPLES / BEST PRACTICES		
10	Display of props or objects (e.g., archaeology, plant, insect or cultural displays)	3,61	Very important
11	Size of the displays or media must reflect life-size examples to relate to the information presented	3,05	Neither important nor less important
12	Displays of all media should be lively/rich/vibrant/exciting	3,36	Neither important nor less important
13	Concise information presented in the media	3,57	Very important
14	Easily readable text or information in relevant media (e.g., not too small font)	3,74	Very important
15	Refrain from technical or academic terms in all media	3,21	Neither important nor less important
16	Staff of the interpretation visitor centre should be able to answer any question related to the information presented	4,15	Very important

Table 2: Continued

	ASPECTS	mean	*Level of importance
17	Opportunity for interpretation visitor centre tours lead by a tour guide	3,36	Neither important nor less important
18	Souvenirs related to interpretation centre can be bought to remember the experience	2,96	Neither important nor less important

* Five-point Likert-type scale ranging from 1 = *Not at all important* to 5 *Extremely important*.

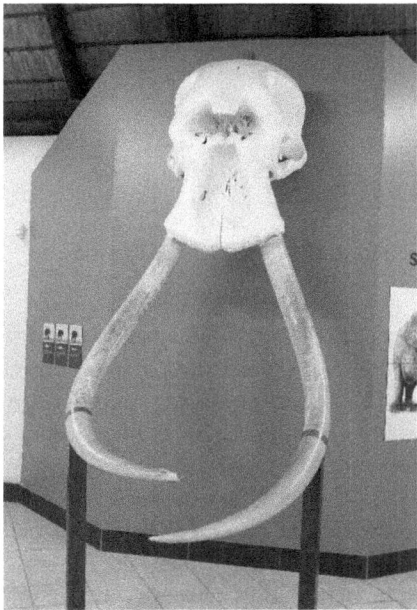

Figure 4: Life-size elephant skull (Photo: P. van der Merwe)

Practical examples

The results in Table 2 clearly indicate that there is a need for an interpretation centre where tourists can spend some time. Tourists expect a centre with direct (i.e., a staff member) and indirect (information boards, props etc.) communication.

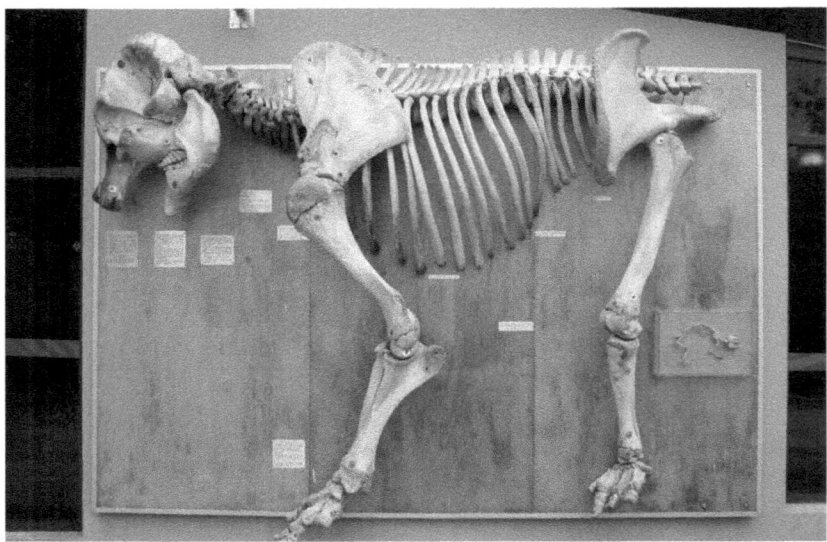

Figure 5: Life-size elephant carcass (Photo: P. van der Merwe)

Tourists also expect the static interpretation services to be multi-sensory: visual and audio. A life-size example of an animal (Figures 4 and 5) that is accompanied by brief, yet interesting, information on the animal and also audio that allows the visitor to listen to the animal's sounds would be an excellent interpretation piece to satisfy these needs. Two-dimensional (2-D) posters can be used in the centre to incorporate textural elements; for example, visitors can feel the difference between the fur of a leopard and a lion, with interesting information on the composition of the fur.

These elements can be designed in a way that is enjoyable for all visitors, as Ham (1992) suggests. The life-size examples of animals can all be incorporated to be part of a jungle gym in which children can have the opportunity to experience, in a playful manner, the relevance of the animal's size compared to theirs. The 2-D posters can also be presented in the form of a game in which the fur needs to be matched with the correct animal or a quiz to test one's knowledge of what was learned previously.

Taking the above beyond a dedicated centre can be achieved by designing a KTP app that contains pictures, information and sounds of the animals that can be used on a game-drive experience in the park. This form of interpretation

Table 3: Agreement and extent of information to be presented

Information should be presented on...	[1]Level of agreement (mean)	[2]Extent of information (mean)
• the fauna of the park	4.64	4.44
• the flora of the park	4.39	4.21
• the ecology of the park	4.36	4.06
• the climatology of the park	4.14	3.87
• the geography of the park	4.13	3.86
• practical matters to conserve the environment	4.13	3.81

1 Measured on a five-point Likert-type scale ranging from 1 *strongly disagree* to 5 = *strongly agree*.
2 Measured on a five-point Likert-type scale ranging from 1 *to a very little extent* to 5 = *to a very large extent*.

limits the information overload a tourist may experience in the interpretation centre, as the tourist is in control of which information they would like to read and learn, and when.

In whatever form, the park can cater for the tourist who wants either limited or detailed information on a matter by using layering: At first, the information can be brief, containing only the most important aspects. If the tourist would like to know more, additional information can be presented by the click of a button in electronic format, providing a QR code that directs the visitors to a website or refers them to another poster.

Information presentation in the information centre

Respondents were asked to indicate what types of information they would prefer in an interpretation centre and to what extent they would like to learn more about the specified topics. Regarding the information types, a five-point Likert-type scale was used (ranging from 1 = *strongly disagree* to 5 = *strongly agree*). A five-point Likert-type scale was also used to determine the extent to which respondents would prefer to obtain information on the various topics (ranging from 1 = *to a very little extent* to 5 = *to a very large extent*). The following forms of information obtained the highest mean values.

It is clear from the information supplied that respondents prefer to learn more about the natural aspects and attractions of the park, with animals, plants and the park's ecology as the most important aspects (Table 3).

Figure 6: Display of different forest insects (Photo: P. van der Merwe)

Furthermore, they would like to learn about these aspects to a greater extent. This information, together with information on other aspects such as the park's climate, geography and aspects that could help conserve the environment, should be presented at the information centre.

Practical examples

Most of the displays in an interpretation centre will be static, but, as previously explained, these displays can be designed to be enjoyable. It may seem difficult to design a display of fauna, flora, ecology, climate and geography to be enjoyable, but if the design is multi-sensory, which tourists expect, it can be accomplished. As the previous examples mostly included fauna, this section will focus on other aspects identified in the results.

The transition from one season to another (climate) and how this affects, for example, flora, can be illustrated in a miniature landscape display of the whole park, where the use of different colours, lights and sound effects can mimic the seasonal change in nature. A voiceover can also be incorporated (which people can listen to by using headphones) to highlight specific geographic areas of importance to the park and how these areas are affected by the seasonal change, or even over a longer period of time. The landscape can include specific landmarks in the display that can refer the visitor to other displays/facilities for more information. One of these displays can relate to specific flora that is unique to the park and can show how it is adapted to survive the winter seasons. Similarly, the relationships between plants and animals (ecology) (Figure 6) during these seasons can also be incorporated with short videos on every season that can be watched separately.

The fact that tourists indicated that they expect information on practical conservation matters amplifies the importance of including such information where the park may need the most support. As discussed in the literature review, this information should be practical in such a way that tourists' behaviour becomes conservation orientated. Considering that conservation can be expensive, a donation box can be of assistance. For example, this box can be designed to illustrate a tree that grows out of the box as the donations increase. Park management can also explore the idea of volunteer work in specific conservation programmes being rewarded in the form of a discount at the restaurant or shop.

3. Conclusions

This chapter provided an analysis of interpretation and how it can be incorporated in protected areas with specific reference to the Kgalagadi Transfrontier Park (KTP), South Africa.

Although ecotourists typically portray the usual traits (for example an appreciation for nature or travelling to relax and to escape daily routine to name but a few), these traits are context specific and it is necessary for any protected area management to determine the needs, expectations and motivations of ecotourists to tailor-make products and services. One of the services that ecotourists expect is high-quality interpretation. Interpretation in protected areas should go beyond the mere presentation of information, and should lead the tourists to some form of pro-conservation behaviour that is in line with the mandate of protected areas. It is against this background that interpretation is used as a 'parks *with* tourists' principle, since it provides benefits to both the tourists and the protected area (increased visitor enjoyment, increased length of stay, increased spending, loyalty, increased visitor awareness and support for conservation practices).

The presented case study, the Kgalagadi Transfrontier Park, highlights the interpretation needs tourists expect from this specific park. Not only were specific interpretation services identified, but the tourists also showed particular interest in specific topics (e.g., climatology or ecology), as well as preferences in specific media in which such interpretation should be presented. This is valuable information for the development of interpretation in the park. Designing interpretation that suits the needs of tourists will ensure satisfied tourists and, in return, will lead to support of the park's conservation practices.

References

Ajzen, I. (1991). The theory of planned behaviour. *Organizational Behaviour and Human Decision Processes*, 50(2), 179–211. doi: https://doi.org/10.1016/0749-5978(91)90020-T

Ballantyne, R., & Hughes, K. (2006). Using front-end and formative evaluation to design and test persuasive bird feeding warning signs. *Tourism Management*, 27(2), 235–246. doi: https://doi.org/10.1016/j.tourman.2004.09.005

Ballantyne, R., Packer, J., Hughes, K., & Dierking, L. (2007). Conservation learning in wildlife tourism settings: lessons from research on zoos and aquariums. *Environmental Education Research*, 13(3), 367–383. doi: https://doi.org/10.1080/13504620701430604

Ballantyne, R., Packer, J., & Hughes, K. (2009). Tourists' support for conservation messages and sustainable management practices in wildlife tourism experiences. *Tourism Management*, 30(5), 658–664. doi: https://doi.org/10.1016/j.tourman.2008.11.003

Ballantyne, R., Packer, J., & Falk, J. (2011). Visitors' learning for environmental sustainability: Testing short- and long-term impacts of wildlife tourism experiences using structural equation modelling. *Tourism Management*, 32(6), 1243–1252. doi: https://doi.org/10.1016/j.tourman.2017.07.015

Bandura, A. (1977). *Social learning theory*. Englewood Cliffs, NJ: Prentice Hall.

Bidder, C., Kibat, S.A., & Fatt, B.S. (2016). Cultural interpretation toward sustainability: a case of Mount Kinabalu. *Procedia: Social and Behavioural Sciences*, 224(2016), 632–639. doi: https://doi.org/10.1016/j.sbspro.2016.05.454

Botha, E., Saayman, M., & Kruger, M. (2016). Expectations versus experience – the Kruger National Park's interpretation services form a regional approach. *Journal of Ecotourism*, 15(2), 158–183. doi: http://dx.doi.org/10.1080/14724049.2016.1178753

Chen, W.Y., & Jim, C.Y. (2012). Contingent valuation of ecotourism development in country parks in the urban shadow. *International Journal of Sustainable Development & World Ecology*, 19(1), 44–53. doi: https://doi.org/10.1080/13504509.2011.588727

Cheung, L.T.O., & Jim, C.Y. (2013). Ecotourism service preference and management in Hong Kong. *International Journal of Sustainable Development & World Ecology, 20*(2), 182–194. doi: ps://doi.org/10.1080/13504509.2013.775192

Cheung, L.T.O., & Jim, C.Y. (2014). Expectations and willingness-to-pay for ecotourism services in Hong Kong's conservation areas. *International Journal of Sustainable Development & World Ecology, 21*(2), 149–149. doi: https://doi.org/10.1080/13504509.2013.859183

De Rojas, C., & Camarero, C. (2008). Visitors' experience, mood and satisfaction in a heritage context: evidence from an interpretation centre. *Tourism Management, 29*(3), 525–537. doi: https://doi.org/10.1016/j.tourman.2007.06.004

De Wet, J.C. (2010). *The art of persuasive communication: a process*. 9th edition. Boston: Allyn and Bacon.

Dioum, B. (1968). Dioum quote. Available at: https://www.goodreads.com/quotes/6430296-in-the-end-we-will-conserve-only-what-we-love Accessed on: 11 December 2018.

Dudley, N., & Stolton, S. (Eds.) (2008). Defining protected areas: an international conference in Almeria, Spain. Gland, Switzerland: IUCN. Available at: <https://portals.iucn.org/library/sites/library/files/documents/2008-106.pdf> Accessed on: 25 May 2018.

Eagles, P.F.J. (2002). Trends in park tourism: economics, finance and management. *Journal of Sustainable Tourism, 10*(2), 132–153. doi: https://doi.org/10.1080/09669580208667158

Eagles, P.F.J. (2014). Fiscal implications of moving to tourism finance for parks: Ontario Provincial Parks. *Managing Leisure, 19*(1), 1–17. doi: https://doi.org/10.1080/13606719.2013.849503

Eunson, B. (2008). *Communicating in the 21st century*. 2nd ed. Milton, Qld.: John Wiley & Sons, Australia.

Festinger, L. (1957). *A theory of cognitive dissonance*. Evanston, Illinois.: Row, Peterson.

Ham, S. (1992). *Environmental interpretation: a practical guide for people with big ideas and small budgets*. Golden, Colorado.: North American Press.

Ham, S., Housego, A., & Weiler, B. (2005). *Tasmanian thematic interpretation planning manual*. Hobart: Tourism Tasmania.

Hughes, M., Ham, S.H., & Brown, T. (2009). Influencing park visitor behaviour: a belief-based approach. *Journal of Park and Recreation Administration, 27*(4), 38–53.

Hwang, S.N., Lee, C., & Chen, H.J. (2005). The relationship among tourists' involvement, place attachment and interpretation satisfaction in Taiwan's national parks. *Tourism Management*, 26(2), 143–156. doi: https://doi.org/10.1016/j.tourman.2003.11.006

IUCN (International Union for Conservation of Nature). (2018). What is a protected area? [online]. Available: <https://www.iucn.org/theme/protected-areas/about> Accessed: 25 May 2018.

Jacobs, M.H., & Harms, M. (2014). Influence of interpretation on conservation intentions of whale tourists. *Tourism Management*, 42(June), 123–131. doi: https://doi.org/10.1016/j.tourman.2013.11.009

Jurdana, D.S. (2009). Specific knowledge for managing ecotourism destinations. *Tourism & Hospitality Management*, 15(2), 267–278.

Kuo, I. (2002). The effectiveness of environmental interpretation at resource-sensitive tourism destinations. *International Journal of Tourism Research*, 4(2), 87–101. doi: https://doi.org/10.1002/jtr.362

Kruger, M., & Saayman, M. (2010). Travel motivation of tourists to Kruger and Tsitsikamma National Parks: a comparative study. *South African Journal of Wildlife Research*, 40(1), 93–102. doi: https://doi.org/10.3957/056.040.0106

Lee, T.H. (2009). A structural model for examining how destination image and interpretation services affect future visitation behavior: a case study of Taiwan's Taomi eco-village. *Journal of Sustainable Tourism*, 17(6), 727–745. doi: https://doi.org/10.1080/09669580902999204

Lück, M. (2003). Education on marine mammal tours as agent for conservation – but do tourists want to be educated? *Ocean & Coastal Management*, 46(9/10): 943 – 956. doi: https://doi.org/10.1016/S0964-5691(03)00071-1

Lück, M. (2015). Education on marine mammal tours – But what do tourists want to learn? *Ocean & Coastal Management*, 103: 25–33. doi: http://dx.doi.org/10.1016/j.ocecoaman.2014.11.002

Marchall, S., Granquist, S.M., & Burns, G.L. (2017). Interpretation in wildlife tourism: assessing the effectives of signage on visitor behaviour at a seal watching site in Iceland. *Journal of Outdoor Recreation and Tourism*, 17(March), 11–19. doi: https://doi.org/10.1016/j.jort.2016.11.001

Maree, K., & Pietersen, J. (2007). Sampling. In K. Maree (Ed.), *First steps in research*. (pp. 171–196). Pretoria: Van Schaik.

Moore, K. (2017). Interpretation needs for Kgalagadi Transfrontier Park [personal interview]. 17 Jan., Kgalagadi Transfrontier Park.

Moscardo, G. (1999). *Making visitors mindful: principles for creating quality sustainable visitor experiences through effective communication*. Champaign, Illinois.: Sagamore.

Moscardo, G., & Ballantyne, R. (2008). Interpretation and attractions. In A. Fyall, B. Garrod, A. Leask & S. Wanhill (Eds.), *Managing visitor attractions* (pp. 237–252). 2nd ed. Oxford: Elsevier.

Moscardo, G., Woods, B., & Saltzer, R. (2004). The role of interpretation in wildlife tourism. In K. Higginbottom (Ed.), *Wildlife Tourism: impacts, management and planning* (pp. 231–251). Common Ground Publishing, Altona, VIC, Australia. Available at: https://researchonline.jcu.edu.au/7500/1/7500_Moscardo_et_al_2004.pdf Date of access: 19 June 2018.

Orams, M. (1994). Creating effective interpretation for managing interaction between tourists and wildlife. *Australian Journal of Environmental Education*, *10*(1), 21–34. doi: https://doi.org/10.1017/S0814062600003062

Orams, M. (1996). A conceptual model of tourist-wildlife interaction: the case for education as a management strategy. *The Australian Geographer*, *27*(1), 39–51. doi: https://doi.org/10.1080/00049189608703156

Pavlov, I. (1958). *Experimental psychology and other essays*. London: P. Owen.

Piaget, J. (1972). *The psychology of intelligence*. London: Routledge & Paul.

Powell, R. B., & Ham, S. H. (2008). Can ecotourism interpretation really lead to pro-conservation knowledge, attitudes and behaviour? Evidence from the Galapagos Islands. *Journal of Sustainable Tourism*, *16*(4), 467–489. doi: https://www.tandfonline.com/doi/full/10.1080/09669580802154223

Sharpley, R. (2006). Ecotourism: a consumption perspective. *Journal of Ecotourism*, *5*(1&2), 7–22. doi: https://doi.org/10.1080/14724040608668444

Sheena, B., Mariapan, M., & Aziz, A. (2015). Characteristics of Malaysian ecotourist segments in Kinabalu Park, Sabah. *Tourism Geographies*, *17*(1), 1–18. doi: https://doi.org/10.1080/14616688.2013.865069

Skinner, B.F. (1935). Two types of pseudo type. *Journal of General Psychology*, *12*(1), 66–77. doi: https://doi.org/10.1080/00221309.1935.9920088

Smith, L.D.G., Curtis, J., Mair, J., & Van Dijk, P.A. (2012). Request for zoo visitors to undertake pro-wildlife behaviour: how many is too many? *Tourism Management*, *33*(6), 1502–1510. doi: https://doi.org/10.1016/j.tourman.2012.02.004

South Africa (International Relations and Cooperation). (2004). *Transfrontier Conservation Areas (TFCAs)* [online]. Available: http://www.dirco.gov.za/foreign/Multilateral/inter/tfcas.htm Date of access: 24 July 2018.

South African National Parks (SANParks). (2018). *Map of Kgalagadi Transfrontier Park* [online]. Available: https://www.sanparks.org/images/parks/kgalagadi/maps/full_parkmap08.jpg Date of access: 31 July 2018.

Stern, P.C. (2000). Toward a coherent theory of environmentally significant behavior. *Journal of Social Issues, 56*(3), 407–424. doi: https://doi.org/10.1111/0022-4537.00175

Stewart, E.J., Hayward, B.M., Devlin, P.J., & Kirby, V.G. (1998). The "place" of interpretation: a new approach to the evaluation of interpretation. *Tourism Management, 19*(3):257–266. doi: https://doi.org/10.1016/S0261-5177(98)00015-6

Tilden, F. (1977). *Interpreting our heritage*. 3rd ed. Chapel Hill: University of North Carolina Press.

Thorndike, E.L. (1911). *Animal intelligence*. New York: Macmillan Co.

United Nations Educational, Scientific and Cultural Organisation (UNESCO). (2017). *The !Xam Khomani Heartland* [online]. Available from: http://whc.unesco.org/en/tentativelists/1910/ Date of access: 24 July 2018.

Ward, C.W. & Wilkinson, A.E. (2006). *Conducting meaningful interpretation: a field guide for success*. Golden Colorado: Fulcrum Publishing.

Weaver, D.B., & Lawton, L.J. (2017). A new visitation paradigm for protected areas. *Tourism Management, 60*(June), 140–146. doi: https://doi.org/10.1016/j.tourman.2016.11.018

Wiener, C.S., Needham, M.D., & Wilkinson, P.F. (2009). Hawaii's real life marine park: interpretation and impacts of commercial marine tourism in the Hawaiian Islands. *Current Issues in Tourism, 12*(5–6), 489–504. doi: https://doi.org/10.1080/13683500902736855

Wijeratne, A.J.C., Van Dijk, P.A., Kirk-Brown, A., & Frost, L. (2014). Rules of engagement: the role of emotional display rules in delivering conservation interpretation in a zoo-based tourism context. *Tourism Management, 42*(June), 149–156. doi: https://doi.org/10.1016/j.tourman.2013.11.012

Woodley, S., Bertzky, B., Crawhall, N., Dudley, N., Londoño, J.M., MacKinnon, K., Redford, K., & Sandwith, T. (2012). Meeting Aichi Target 11: What does success look like for protected areas? *Parks, 18*(1), 23–36. Available from: https://cmsdata.iucn.org/downloads/parks_woodley_1.pdf Date of access: 27 June 2018.

Zeppel, H., & Muloin, S. (2008). Conservation benefits of interpretation on marine wildlife tours. *Human Dimensions of Wildlife, 13*(4), 280–294. doi: https://doi.org/10.1080/10871200802187105

Chantal D. Pagel, Matthias Waltert, Michael Scheer and Michael Lück

Swimming with wild orcas in Norway: Killer whale behaviours addressed towards snorkelers and divers in an unregulated whale watching market

1. Introduction

Orcinus orca is an incredibly powerful and capable creature, exquisitely self-controlled and aware of the world around it, a being possessed of a zest for life and a healthy sense of humor, and moreover, a remarkable fondness for and interest in humans. (Paul Spong, *Mind in the Waters*, 1974)

The unique relationship between humans and killer whales is deeply rooted in spiritual connections and plays a dominant role in North West Pacific culture and folklore (Knudtson, 1996). The appreciation for orcas started long before the 1960s, when killer whales on public display shaped public perceptions in a positive way (Knudtson, 1996), and the early 1990s, when the motion picture *Free Willy* influenced a whole generation. Notably, their natural curiosity about humans and highly cooperative social structures have contributed tremendously to the popularity this species is experiencing today (Frohoff & Peterson, 2003). Therefore, it is less surprising that killer whales are one of the best-studied marine mammals worldwide (Le Duc, Robertson, & Pitman, 2008). Research on orcas to date has contributed to a profound knowledge of cetacean behaviour, ecology, social complexity and acoustics (e.g., Baird & Stacey, 1988; Bigg, Olesiuk, Ellis, Ford, & Balcomb, 1990; Ford, 1989; Williams, Trites, & Bain, 2002).

With their rising popularity, there has been correlated growth in the demand for seeing orcas in their natural habitat, which has contributed significantly to today's reign of marine wildlife tourism and whale-watching in particular (Hoyt, 2012). The demand for closer and more personal interactions with marine wildlife has been catered for by the commercialisation of in-water interactions (swim-with programmes; SWPs) with dolphins, both in human care and in the wild (Samuels, Bejder, & Heinrich, 2000). Norway is the only country in the world where commercial and unregulated SWPs with orcas have been regularly offered on a small scale. However, with this opportunity there also comes the need to understand how these animals react to close approaches (Pagel, Scheer, & Lück, 2017). Killer whales are highly mobile predators and, where potentially dangerous animals are involved, humans are at risk of experiencing aggressive, defensive or dangerous play behaviour (Newsome & Rodger, 2013). Such situations may increase when tour participants are unfamiliar with orcas' natural responses, which is why baseline research on animal behaviour prior to the establishment of commercial in-water interactions is pivotal.

2. Swim-with programmes and cetacean behaviour

The shift from non-invasive, passive observations and the feeling of being 'barely there' (Hoyt, 2003) to close, interactive encounters can explain the emergence of SWPs as a new phenomenon with a wide distribution, globally as well as within the animal kingdom. Interactions are offered with individuals or groups of animals as well as with whole populations of marine mammals. Accessibility and a predictable distribution of marine wildlife both play a major role for these ventures, which are generally offered with species that show non-aggressive behaviours in areas that are considered to be safe for participants (United Nations Environment Programme/Conservation of Migratory Species, 2017). However, a substantial number of commercial tours operating in open-water environments target wildlife that can potentially pose a threat to human health and safety.

For many, swimming with cetaceans describes a lifelong ambition and its fulfilment requires a comprehensive management approach to ensure tourist health and safety as well as minimal impact on the targeted population (Orams, 1997). However, the management of cetacean-focused tourism must address multiple domains (ecological, social, political, economic and cultural; Lundquist, 2014), highlighting its complexity. Negative ecological outputs have been studied extensively within the cetacean-based tourism phenomenon (Higham, Bejder, & Williams, 2014), but little is known about the initiated behaviour of cetaceans towards swimmers. Whales and dolphins are considered to be highly charismatic animals and, as such, induce positive emotions in human observers (Kellert, 1987). Due to this, dolphins still command the popular image of being playful and overall safe animals to interact with (Orams, 1997) despite the fact that intraspecifically and during mixed-species associations they display aggression as part of their behavioural makeup, invalidating the common view of dolphins being friendly and passive (Herzing & Elliser, 2013; Herzing & Johnson, 1997; Ridgway, 1990). Wild dolphins display behaviours towards human swimmers that they also use during intraspecific interactions (e.g. Dudzinski, 1996; Herzing, 2006; Scheer, 2010; Shane, 1990). And non-affiliative behaviours cannot be categorically excluded as potential elements during in-water encounters. Non-affiliative responses to swimmers have been observed predominantly among habituated, solitary and sociable, food-provisioned and captive individuals (Orams, Hill, & Baglioni, 1996; Samuels et al., 2000). Therefore, it is often assumed that interactions with unhabituated animals provide a low risk for encountering threatening behaviours (Perrine, 1998; Samuels et al., 2000), ignoring that cetaceans are large, powerful predators that, as with any wildlife, may behave unpredictably and are capable of injuring or even killing people (Orams, 1997). A life-threatening incident involving a female swimmer who was pulled underwater by an unhabituated, male short-finned pilot whale (*Globicephala macrorhynchus*) in Hawaiian waters provides evidence that any encounter with wild cetaceans can be dangerous (Shane, 1995; Shane, Tepley, & Costello, 1993). However, it has to be stressed that the incident occurred after the swimmer initiated physical contact with the individual and substantiates the significance of educational management mechanisms regarding SWPs.

Negative responses also can occur due to humans not taking natural responses into account (Shackley, 1996). Swimmers who have little experience with encountering marine mega fauna are especially prone to misinterpreting behavioural patterns and are also often unaware of behaviours of their own that could provoke undesirable responses from wildlife (Pagel, Scheer, & Lück, 2018). For instance, during close approaches, behaviours being expressed due to stress (e.g. tail slaps, head nod, open mouth squeak) have been misinterpreted by swimmers as signs of enjoyment, which has resulted in agonistic behaviours from the dolphins, with swimmers being bitten and charged (Doye, 1995; Dudzinski, 2003; Frohoff, 2003). Dolphins use their whole bodies (including their flukes, pectoral fins, teeth as well as their rostrum), movements and vocalisations for communicative purposes, revealing episodes of affection, sexual overture and aggression (Orams et al., 1996). Agonistic behaviours can be found across species and include jerky movements of extremities or the whole body, an s-shaped posture or arching back, spread fins, jaw clapping or head shakes with an open mouth as well as pushy behaviour (Herzing, 2006; Mann, Connor, Tyack, & Whitehead, 2000; Norris et al., 1985; Orams et al., 1996; Overstrom, 1983; Scheer, Hofmann, & Behr, 2004). An animal's body language can thus reliably provide clues ('warning signs') of a prefaced aggression or even an attack. For swimmers, sexual behaviour from an animal might become additionally dangerous. All these behaviours are familiar to professional observers;

however, they can be overlooked by untrained and inexperienced swimmers (Stringham, 2011).

The establishment of behavioural catalogues can provide valuable insights into a species' behavioural repertoire and have been developed by field biologists for a range of terrestrial taxa (Lehner, 1987). Such ethograms also cover some odontocete species, such as Yangtze finless porpoises (*Neophocaena phocaenoides asiaeorientalis*; Xian, Wang, Jiang, Zheng, & Wang, 2010), short-finned pilot whales (Scheer et al., 2004), captive as well as wild, sociable belugas (*Delphinapterus leucas*; Campbell, 2011; Frohoff, Kinsman, Rose, & Sheppard, 2000), and for Amazon botos (Scheer, de Sá Alves, Ritter, Azevedo, & Andriolo, 2014).

The first ethogram for killer whales was elaborated by Martinez and Klinghammer in 1978 and aimed to contribute to a standardised behavioural terminology for this species. Behavioural observations were made on free-ranging animals in the Pacific North West as well as on seven individuals held in human care, demonstrating a large repertoire of over 50 behavioural patterns during interactions with conspecifics as well as with other cetacean species. Here, head bobbing and open mouth orientation towards conspecifics or other species were identified as aggressive behaviours. Further, breaching as well as fluke shakes/slaps were recognised as threat displays; however, the latter patterns can also be interpreted as non-aggressive communication (e.g., during foraging), highlighting the need to interpret animal behaviour in context (Dudzinski, 1996; Stringham, 2011).

The importance of such behavioural studies was outlined by Scheer (2010), as to date there is no widely accepted behavioural catalogue of human–cetacean encounters in use among researchers.

3. The development of killer whale tourism in Norway

In Norway, commercial killer whale safaris ('Spekkhogger Safaris') began in 1985 in the Lofoten area, followed by the development of operations in neighbouring Tysfjord where the boat, Øyprinsen, took the first tourists out to see whales in the autumn of 1992 (Damsgård, 2000; Stenersen & Similä, 2004). The development of killer whale safaris was a direct response to the seasonal occurrence of an abundance of orcas foraging on the herring that overwinter in the fjords from October to January (Similä, Holst, & Christensen, 1996). The large numbers of orca enabled these ventures to quickly become a huge success, providing Norway the ability to offer a world-class nature-based wildlife experience (Stenersen & Similä, 2004). Over time, wildlife photographers and journalists fuelled public awareness by providing spectacular images and footage of this nature event in the Arctic, resulting in more and more people wanting to encounter killer whales up close, which still continues to date (Pagel et al., 2018; Stenersen & Similä, 2004). As a consequence, the Norwegian whale-watching market became flooded with operators but also with recreational boaters using smaller, highly mobile craft. With the latter whale watchers, more aggressive boat handling was witnessed, which prompted the setting up a code of conduct, bringing operators, tourists, NGOs and scientists together to attempt a solution (Damsgård, 2000; Stenersen & Similä, 2004). Despite the efforts of the past 20 years, the Norwegian whale-watching market is still unregulated (Pagel et al., 2018), not least due to rivalry and mistrust among local stakeholder groups representing different interests (Stenersen & Similä, 2004).

Although orca safaris have been established for quite some time, SWPs with killer whales only started to proliferate at the start of the new millennium (Damsgård, 2000; Stenersen & Similä, 2004).

4. Interactive behavioural responses of Norwegian killer whales

In the early 2000s, when commercial SWPs with killer whales were still in their infancy, Stenersen and Similä (2004) had already pointed to the importance of undertaking baseline research into understanding orca behaviour before letting people entering the water on a regular basis. Yet this proposed area of research was never taken up, even when the numbers of participants began to grow. Almost 40 years after the development of the first ethogram describing general killer whale behaviours, and after 15 years of unregulated swimming operations in northern Norway, with most information on behaviour being anecdotal, the first ethogram on the self-initiated, interactive behaviour of killer whales was finally established (Pagel, 2015; Pagel et al., 2017).

This study contributes significantly to the understanding of behaviours that can be encountered during close interactions with these animals and describes the first research on SWPs targeting Norwegian killer whales. The scientific determination of behavioural categories was achieved through the application of video-based underwater data analysis using videos provided by wildlife filmmakers as well as empirically collected footage sampled in Tysfjord, Vestfjord and Hamn i Senja between 2000 and 2015 (Figure 1).

In total, 58 video clips of human–orca interactions were examined with a total length of 27 minutes and 32 seconds. Footage varied greatly in length (0:08 to 19:14 minutes) depending on whether clips were edited (mean length 2.5

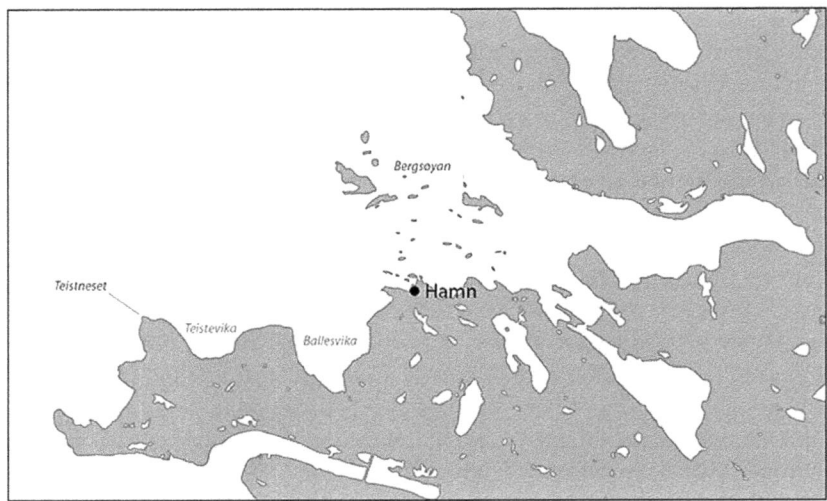

Figure 1. Study Area

minutes). Two video clips (numbers 57 and 58; see Table 1) were collected by the first author during fieldwork in January 2015. The research base was located in the village of Hamn on Norway's second largest island, Senja, which is situated within the Arctic Circle between Andøya and Tromsø. The island has protected fjords and islets that provide good conditions for field research. The target population were killer whale groups following the spring-spawning herring to this area during winter months (October to February), which had been observed during the three seasons prior to the research season, 2014/2015. Killer whale groups were encountered in an area between Hamn harbour, Bergsøyan, the bays Ballesvika and to the western tip of Teistevika, Teistneset, in relatively shallow waters (from 25 metres in Ballesvika and Teistevika to 150 metres outside the bays).

Since the killer whale groups found in this area are high in numbers and typically migrate from other fjords to feed on herring, it is likely that the population is unhabituated to swim activities. However, some individuals were identified from the previous winter season, so it can be assumed that at least some animals return to their wintering grounds (Similä et al., 1996; T. Similä, personal communication, February, 15, 2015).

Killer whale groups were approached with a motorised 8m long sailboat at a speed of between one and six knots when the animals were displaying various behavioural states, with feeding, travelling and resting being the most frequently observed. Swimmer placement was conducted at a distance of two to ten metres, given that the animals showed an interest in the boat and allowed the boat nearby. When animals showed avoidance behaviour such as diving away, increased swim speed or sudden changes in direction, approaching the target group was abandoned.

To ensure a less invasive approach during data collection, a precautionary protocol, following the guidelines of the Undersea Soft Encounter Alliance (USEA), was applied. This included:

- Swimmers entered the water as quietly as possible and avoided splashing.
- As soon as animals were in sight, forward swimming was stopped and swimmers remained motionless and passive on the water's surface.
- When animals departed from the area, the swim attempt was terminated.
- Any initiation of physical contact was avoided.

The first author and the additional filmmaker were both experienced snorkelers and certified scuba divers. They had previously visited the study area several times. During encounters, they were wearing snorkelling gear and drysuits. Snorkelling, as an approach method, had the advantages of quick preparation on board and less irritation to the orcas, as scuba diving might have affected the animals' behaviour resulting in avoidance (Scheer et al., 2004). Particularly in odontocete species, the display of bubbles is considered to be a means of communication (Pryor, 1990); consequently, air bubbles from scuba gear could influence the animals' behaviour. The encounters were recorded with an action camcorder on a floating pole to facilitate camera handling.

All footage was initially inspected for interactive behaviours using the *ad libitum* method (Altmann, 1974; Martin & Bateson, 1993). Interactive behaviours were defined as behaviours initiated by an individual or a group of killer whales, and directed towards a swimmer within a human body length (< 2 m) to 20 m range. Based on Scheer (2010), interactive behaviours were categorised as:

a. 'affiliative/neutral' when there were no signs of threat or aggression;
b. 'aggression/threat' when they put swimmers at physical risk and/or the same behaviour was observed by other researchers during agonistic intraspecific interactions among killer whales or other toothed whale species (Dudzinski, 1996; Martinez & Klinghammer, 1978; Shane, 1990); or
c. 'sexual' when sexual behaviours were directed towards swimmers, putting them at physical risk.

After establishing an *á priori* ethogram containing all behaviours detected during the initial inspection, all videos were analysed a second time.

Another commonality with the study conducted by Scheer et al. (2004, 2014) was the use of a 'one/zero' sampling method, which has been described as reliable and the only practicable method for recording intermittent behaviour when recording sessions have been divided into short sampling intervals (Martin & Bateson, 1993). During an interval, the observer records whether a particular behavioural pattern has occurred, regardless of its continuance and frequency. The 'rate' in the one/zero sampling method is calculated as the proportion of all sample intervals during which the behavioural pattern occurred (Martin & Bateson, 1993).

Using this method, when a specific behavioural pattern was observed in a video recording, it was labelled as 'x' whereas an absence was indicated with a dash (-). The score (x or -) was entered onto a table beside the video number and beneath the code (abbreviation) of the interspecific behaviour pattern observed (Table 1).

The completion of the ethogram on Norwegian killer whales revealed a variety of interactive behaviours directed towards humans during in-water encounters. The ethogram was established out of eight interspecific behavioural patterns, which were all considered to be affiliative in nature as no aggressive or sexual behaviours were found. The rate of one/zero occurrences for each of the eight behaviours varied from 0.07 to 0.74.

The most common behaviours during encounters with human swimmers and divers were whistling/calling (Figure 2), occurring in 43 out of the 58 clips (0.74) and eye contact (Figure 3), found in 39 videos (0.67). Bubble release (Figure 4) and encircling (Figure 5) were least common and only observed in four (0.07) and six video clips (0.10), respectively. Echolocation (Figure 6; 0.62), close approach (Figure 7; 0.48) and head orientation (Figure 8; 0.38) were also common patterns, whereas belly-up approaches to swimmers (Figure 9; 0.21) were recognised sporadically.

Of the eleven affiliative, interspecific behaviours found in short-finned pilot whales described by Scheer et al. (2004, 2014), six were also found in killer whales; whereas the agonistic behaviours, such as the reported head shakes in pilot whales or threat displays (e.g. approach with an open mouth), could not be confirmed for killer whales. However, aggressive head orientation and open mouth posturing are part of the behavioural repertoire of killer whales, as confirmed by Martinez and Klinghammer (1978). The absence of life-threatening patterns might be explained by the swimmers and divers in the study not initiating any physical contact with the animals. Touching can act as a catalyst

Table 1. One/zero (x/-) occurrence of interspecific behaviours of Norwegian killer whales during human swimmer encounters (after Scheer et al., 2014)

Video clip no.	Behaviour pattern*							
	BA	BR	CA	ELO	ENC	EC	HO	W/C
1	-	-	-	x	-	-	x	x
2	-	-	-	x	-	x	x	-
3	-	-	-	x	-	x	x	x
4	-	-	-	-	-	-	-	x
5	-	-	-	x	-	-	-	x
6	x	-	x	x	-	x	x	-
7	x	-	x	x	-	x	x	x
8	-	-	x	x	-	x	x	-
9	-	-	-	-	-	-	-	x
10	-	-	-	-	-	x	x	-
11	x	-	x	-	-	x	-	-
12	-	-	-	-	-	-	-	-
13	-	-	-	x	-	x	x	x
14	x	-	-	-	-	-	x	x
15	-	-	-	-	-	-	x	x
16	x	-	x	x	x	x	x	x
17	x	-	x	x	x	x	x	x
18	x	-	x	x	x	x	x	x
19	-	-	x	x	x	x	-	x
20	-	-	x	x	-	x	x	-
21	-	-	-	-	-	x	-	x
22	-	-	-	x	-	x	x	-
23	-	-	-	-	-	-	-	-
24	-	-	-	-	-	-	-	-
25	-	-	-	x	-	-	-	x
26	-	-	-	x	-	-	-	x
27	x	-	x	-	-	x	x	-
28	-	-	x	-	-	x	-	x
29	-	-	-	-	-	x	-	-
30	-	-	-	-	-	-	-	x
31	-	-	x	x	-	-	-	x
32	-	-	x	x	x	x	-	x
33	-	x	x	x	x	x	x	x

(continued on next page)

Table 1. Continued

Video clip no.	Behaviour pattern*							
	BA	BR	CA	ELO	ENC	EC	HO	W/C
34	x	-	x	x	-	x	x	x
35	-	-	-	x	-	-	-	x
36	x	-	x	x	-	x	-	x
37	-	-	x	x	-	x	-	x
38	-	x	x	x	-	x	-	x
39	-	x	-	x	-	-	-	x
40	x	-	x	x	-	x	-	x
41	-	-	-	x	-	-	-	x
42	-	-	-	x	-	x	-	x
43	-	-	x	x	-	x	-	x
44	-	-	x	-	-	x	-	-
45	-	-	-	-	-	-	-	x
46	-	-	-	x	-	x	-	x
47	-	-	-	x	-	x	x	x
48	-	-	-	x	-	-	-	x
49	-	-	-	x	-	x	-	x
50	-	-	-	-	-	-	-	x
51	-	-	x	-	-	x	x	-
52	-	-	x	-	-	x	-	-
53	-	-	x	-	-	x	-	x
54	x	-	x	x	-	x	-	x
55	-	-	x	x	-	x	-	x
56	-	x	x	x	-	x	x	x
57	-	-	x	-	-	x	x	x
58	-	-	-	-	-	x	-	x
Total	12	4	28	36	6	39	22	43

*BA = belly-up approach; BR = bubble release; CA = close approach; ELO = echolocation; ENC = encircling; EC = eye contact; HO = head orientation; W/C = whistling/calling

for aggressive interactions, which has been discussed by Scheer (2010) and is supported by findings in research addressing other odontocete species during in-water interactions (e.g. Goffman, Lavalli, Kerem, & Spanier, 1999; Mann & Smuts, 1999; Shane et al., 1993). Therefore, avoiding any physical contact with wild killer whales is recommended in order to minimise the risk of generating dangerous behaviour.

Figure 2. Whistling/Calling

Figure 3. Eye Contact

Figure 4. Bubble Release

Figure 5. Encircling

Figure 6. Echolocation

Figure 7. Close Approach

The confirmation of eye contact as one of the most frequent behaviours is

Figure 8. Head Orientation

Figure 9. Belly up Approach

significant as it plays a critical role during commercial SWPs (Curtin, 2006). In general, cetaceans readily make eye contact with humans as well as with each other; however, this is usually brief. Unlike various terrestrial mammals, such as primates, where the dominant individual stares and the submissive animal looks away first, dominant cetaceans are the first to look away during intraspecific interactions (Pryor, 1990). Cetacean eyes can also reveal intentions or internal states; for example, the sclerae of killer whales turn red when expressing aggressive behaviour (Defran & Pryor, 1980). As cetaceans are monocular laterally but possess binocular vision downward and to some extent also to the rear (Mass & Supin, 2009; Pryor, 1990), this can be an explanation for the close approaches with belly presentation found in killer whales. However, other scholars observed belly ups during play, foraging, agonistic and mating behaviours but point out that it is vital to place animal behaviour into context to clarify meaning and function (Mann et al., 2000; Stringham, 2011). Belly ups as a friendly greeting gesture were also observed among Hawaiian spinner dolphins (*Stenella longirostris*) which tilt sidewise to flash their white belly to conspecifics (Norris et al., 1985).

Norwegian killer whales seem to be less receptive to encounters with humans, but simultaneously do not seem to be distracted by human presence from foraging in large herring shoals, which is particularly attractive for tourists who want to film and photograph the animals behaving naturally.

5. Management and future prospects

Today, the number of people seeking exceptional tourism experiences is steadily growing, confirming Shackley's (2001) prediction of tourists demanding greater opportunities to interact with wildlife. The demand is supported by social media content displaying close human–wildlife encounters – images that are shaping the interposition that wildlife is readily accessible and creating expectations about encounters that are often unrealistic (Spradlin, Barre, Lewandowski, & Nitta, 2001; Newsome & Rodger, 2013).

The opportunity to swim with wild killer whales in Norway is unrivalled worldwide due to Norway's available commercial infrastructure and the absence of legislation that prohibits SWPs. It has gained more attention over recent years due to wildlife filmmakers and photographers showing off their content on various social media platforms, fuelling many people's desire to experience the seemingly intimate and prolonged interactions themselves. These expectations can be misguided, and disappointment ensues when encounters are brief and the environment is challenging due to currents, swell, limited underwater visibility, low water temperatures, limited daylight hours as well as swimmers' own skills (Curtin, 2006; Pagel et al., 2018). Apart from the skilled photographers, swimmers are potentially inexperienced snorkellers who need more guidance during the operations (Pagel et al., 2017). However, in terms of the present study, it is essential that all kinds of wildlife tourists who wish to get close to killer whales, whether they are specialised or novice participants, have an understanding of the unpredictable and uncontrollable killer whale behaviours that can be encountered during SWPs in the open ocean. Though agonistic behaviours were not found during this research, this result has to be treated with caution as the availability of relevant footage was limited and further investigation might have revealed additional behaviours that could have been threatening for human swimmers and divers.

The management of tourist behaviour is a critical aspect of safety in wildlife tourism ventures. The 'look but do not touch' policy must be followed at all times as topic-related literature has furnished evidence that threatening behaviours towards humans displayed by free-ranging cetaceans is often the result of swimmers seeking physical contact. To date, there has been no known attack on humans by a wild killer whale. However, encounters were comparatively rare until recently with the noticeable increase in participants in commercial SWPs in Norway, as well as increased opportunistic encounters between killer whales and recreationists (e.g. in New Zealand). These unstructured swims in sometimes challenging open-water environments generate uncertainty in how operators and participants should behave, raising questions of how close one should get (Garrod, 2008). The risk of injury to both humans and wildlife, due in part to close-range boat manoeuvers for placing swimmers as near to the animals as possible, have been pointed out as key problems of SWPs. Yet it is a poorly addressed issue, and so far no regulation or code of conduct takes these issues into account (Allen, 2014; Dans, Crespo, & Coscarella, 2017). Further, the lack of information about negative impacts resulting from inappropriate operator behaviour in providing for recreationists describes a significant management gap (Pagel et al., 2018). Thus, this study provides key findings for future management, supports long-term monitoring, and delivers data for interpretive actions. It is understood that there is no 'one size fits all' solution when it comes to the management regimes of SWPs in different countries; however, it is always advisable to take the human component of these ventures into account, as management approaches are tourist rather than wildlife-oriented (Orams, 2000). Scholars have noted that SWPs might be "a step too far in the portfolio of the wildlife tourism product" (Curtin & Garrod, 2008, p. 110) with close-up personal encounters deemed inappropriate for some species (Orams, 2000). However, as SWPs with marine mega fauna have gained much public attention (including in Norway), and are expected to increase in popularity over the coming years, a recently suggested alternative is to encourage customers to follow animal welfare and conservation strategies and to foster ethical consumption (Moorhouse, D'Cruze, & Macdonald, 2017). Therefore, it is strongly advised that tour participants are taught about orca behaviours prior to their in-water encounters, to empower them with the ability to choose whether to terminate or to continue their encounter accordingly. Tour operators should respect approach distances by not placing swimmers in the animals' path of travel and should be able to identify critical behaviours. Further, through their public outreach, NGOs as key stakeholders can act as mediators by accepting wildlife tourism as an opportunity for nature conservation and education of the wider public (Green, 2017) and can guide tourist decision-making towards selecting more responsible and certified operators.

Although the situation has become noticeably more unstructured in the Tromsø area since the fieldwork was conducted (S. Gust, personal communication, November 27, 2017), the fractional character of the local infrastructure and the seasonality of the killer whale presence in Norwegian waters, as well as limited daylight hours and unpredictable weather conditions, are market-limiting factors that diminish the probability of the development of mass-tourism. Further, the total number of animals visiting the wintering grounds support the assumption that potential tourism pressure is equally distributed and not focused on a single pod, as is the case in the northern and southern resident killer whale populations in the Pacific North West (Pagel, 2015).

The collection of baseline information to investigate the long-term effects of human–wildlife interactions is still in its early stages due to the fast-growing nature of wildlife-oriented tourism. The present results facilitate the general interpretation of interspecific killer whale behaviour, which is essential for safer interactions with these marine predators and enables a sound comparison with similar studies featuring other odontocetes.

References

Allen, S. J. (2014). From exploitation to adoration: The historical and contemporary contexts of human-cetacean interactions. In J. Higham, L. Bejder & R. Williams (Eds.), *Whale-watching: Sustainable tourism and ecological management* (pp. 31–47). Cambridge, England: Cambridge University Press.

Altmann, J. (1974). Observational study of behaviour: Sampling methods. *Behaviour, 49*, 227–267.

Baird, R.W., & Stacey, P.J. (1988). Variation in saddle patch pigmentation in populations of killer whales (*Orcinus orca*) from British Columbia, Alaska, and Washington State. *Canadian Journal of Zoology, 66*, 2582–2585.

Bigg, M.A., Olesiuk, P.F., Ellis, G.M., Ford, J.K.B., & Balcomb, K.C. (1990). Social organization and genealogy of resident killer whales (*Orcinus orca*) in the coastal waters of British Columbia and Washington State. *International Whaling Committee Special Issue, 12*, 383–405.

Campbell, C.A. (2011). Comprehensive ethogram on beluga (*Delphinapterus leucas*) behaviour (Unpublished master's thesis). Texas A&M University, TX.

Curtin, S. (2006). Swimming with dolphins: A phenomenological exploration of tourist recollections. *International Journal of Tourism Research, 8*, 301–315.

Curtin, S., & Garrod, B. (2008). Vulnerability of marine mammals to diving tourism activities. In B. Garrod & S. Goessling (Eds.), *New frontiers in marine tourism: Diving experiences, sustainability, management* (pp. 93–113). Amsterdam, The Netherlands: Elsevier.

Damsgård, B. (2000). Hvalsafari. In *Spekkhogger – Staurkval. Ottar 230/2000*. Tromsø, Norway: Tromsø Museum, Universitetsmuseet.

Dans, S.L., Crespo, E.A., & Coscarella, M.A. (2017). Wildlife tourism: Underwater behavioral responses of South American sea lions to swimmers. *Applied Animal Behaviour Science, 188*, 91–96. https://doi.org/10.1016/j.applanim.2016.12.010

Defran, R.H., & Pryor, K. (1980). The behavior and training of cetaceans in captivity. In L.M. Herman (Ed.), *Cetacean behavior: Mechanisms and functions* (pp. 319–362). New York, NY: J. Wiley.

Doye, S. (1995). *To touch the magic*. Draft of a thesis submitted for degree of Master of Arts, Australian National University, Canberra, Australia.

Dudzinski, K.M. (1996). Communication and behaviour in the Atlantic spotted dolphins (*Stenella frontalis*): Relationships between vocal and behavioural activities (Unpublished doctoral thesis). Texas A&M University, TX.

Dudzinski, K.M. (2003). Letting dolphins speak: Are we listening? In T. Frohoff & B. Peterson (Eds.), *Between species: Celebrating the dolphin-human bond* (pp. 286–295). San Francisco, CA: Sierra Club Books.

Ford, J.K.B. (1989). Acoustic behaviour of resident killer whales (*Orcinus orca*) in British Columbia. *Canadian Journal of Zoology, 67*, 727–745.

Frohoff, T.G. (2003). The kindred wild. In T. Frohoff & B. Peterson (Eds.), *Between species: Celebrating the dolphin-human bond* (pp. 56–71). San Francisco, CA: Sierra Club Books.

Frohoff, T.G., Kinsman, C., Rose, N.A., & Sheppard, K. (2000). *Preliminary study of the behavior and management of solitary, sociable white whales (Delphinapterus leucas) in Eastern Canada* (Report No. SC/52/WW). Cambridge, England: International Whaling Commission Scientific Committee.

Frohoff, T., & Peterson, B. (2003). *Between species: Celebrating the dolphin-human bond*. San Francisco, CA: Sierra Club Books.

Garrod, B. (2008). Marine wildlife tourism and ethics. In J. Higam & M. Lück (Eds.), *Marine wildlife and tourism management* (pp. 257–271). Cambridge, MA: CABI.

Goffman, O., Lavalli, K., Kerem, D., & Spanier, E. (1999). Consequences of swimming with a lone female bottlenose dolphin (*T. aduncus*) in the Gulf of Eilat/Aqaba, Red Sea. In K.M. Dudzinski, T.G. Frohoff & T.R. Spradlin (Eds.), *Wild dolphin swim program workshop proceedings*. Maui, HI: The Society of Marine Mammalogy.

Green, R.J. (2017). Disturbing Skippy on tour: Does it really matter? Ecological and ethical implications of disturbing wildlife. In I. Borges de Lima &

R.J. Green (Eds.), *Wildlife tourism, environmental learning and ethical encounters: Ecological and conservation aspects* (pp. 221–234). Cham, Switzerland: mSpringer.

Herzing, D.L. (2006). The currency of cognition: Assessing tools, techniques, and media for complex behavioural analysis. *Aquatic Mammals, 32*(4), 544–553.

Herzing, D.L., & Elliser, C.R. (2013). Directionality of sexual activities during mixed-species encounters between Atlantic spotted dolphins (*Stenella frontalis*) and bottlenose dolphins (*Tursiops truncatus*). *International Journal of Comparative Psychology, 26*, 124–134.

Herzing, D.L., & Johnson, C.M. (1997). Interspecific interactions between Atlantic spotted dolphins (*Stenella frontalis*) and bottlenose dolphins (*Tursiops truncatus*) in the Bahamas, 1985–1995. *Aquatic Mammals, 23*, 85–99.

Higham, J., Bejder, L., & Williams, R. (2014). *Whale-watching: Sustainable tourism and ecological management.* Cambridge, England: Cambridge University Press.

Hoyt, E. (2003). Toward a new ethic for watching dolphins and whales. In T. Frohoff & B. Peterson (Eds.), *Between species: Celebrating the dolphin-human bond* (pp. 169–177). San Francisco, CA: Sierra Club Books.

Hoyt, E. (2012). *Whale watching blueprint: I. Setting up a marine ecotourism operation.* North Berwick, Scotland: Nature Editions.

Kellert, S.R. (1987). The educational potential of the zoo and its visitor. *Philadelphia Zoo Review, 3*(1), 7–13.

Knudtson, P. (1996). *Orca: Visions of the killer whale.* San Francisco, CA: Sierra Club Books.

Le Duc, G.R, Robertson, K.M., & Pitman, R.L. (2008). Mitochondrial sequence divergence among Antarctic killer whale ecotypes is consistent with multiple species. *Biological Letters, 4*(4), 426–429.

Lehner, P.N. (1987). Design and execution of animal behaviour research: An overview. *Journal of Animal Science, 65*, 1213–1219.

Lundquist, D. (2014). Management of dusky dolphin tourism at Kaikoura, New Zealand. In J. Higham, L. Bejder & R. Williams (Eds.), *Whale-watching: Sustainable tourism and ecological management* (pp. 337–351). Cambridge, England: Cambridge University Press.

Mann, J., Connor, R.C., Tyack, P.L., & Whitehead, H. (2000). *Cetacean societies: Field studies of dolphins and whales.* Chicago, IL: The University of Chicago Press.

Mann, J., & Smuts, B. (1999). Behavioural development in wild bottlenose dolphin newborns (*Tursiops sp.*). *Behaviour, 136*, 529–566.

Martin, P., & Bateson, P. (1993). *Measuring behaviour: An introductory guide* (2nd ed.). Cambridge, England: Cambridge University Press.

Martinez, D.R., & Klinghammer, E. (1978). A partial ethogram of the killer whale. *Carnivore, 1*, 13–27.

Mass, A.M., & Supin, A.Y. (2009). Vision: Retinal topography and visual field organization: Cetaceans. In W. Perrin, W. Würsig & J.G.M. Thewissen (Eds.), *Encyclopedia of marine mammals* (p. 1208). Burlington, MA: Academic Press.

Moorhouse, T., D'Cruze, N.C., & Macdonald, D.W. (2017). Unethical use of wildlife in tourism: What's the problem, who is responsible, and what can be done? *Journal of Sustainable Tourism, 25*, 505–516. doi:10.1080/09669582.2016.1223087

Newsome, D., & Rodger, K. (2013). Wildlife tourism. In A. Holden & D. Fennell (Eds.), *The Routledge handbook of tourism and the environment* (pp. 345–358). New York, NY: Routledge.

Norris, K.S., Würsig, B., Wells, R.S., Würsig, M., Brownlee, S.M., Johnson, C., & Solow, J. (1985). The behavior of the Hawaiian spinner dolphin, *Stenella longirostris* (Administrative report LJ-85-06C). La Jolla, CA: National Marine Fisheries Service, Southwest Fisheries Center.

Orams, M.B. (1997). Wild dolphin based tourism: Minimizing the risks and maximizing the benefits. In N.K. Saxena (Ed.), *Recent advances in marine science and technology*, 96 (pp. 477–489). Honolulu, HI: Pacific Congress on Marine Science and Technology.

Orams, M.B. (2000). Tourists getting close to whales, is it what whale watching is all about? *Tourism Management, 21*(6), 561–569.

Orams, M.B., Hill, G.J., & Baglioni, A.J. (1996). 'Pushy' behavior in a wild dolphin feeding program at Tangalooma, Australia. *Marine Mammal Science, 12*, 107–117.

Overstrom, N.A. (1983). Association between burst-pulse sounds and aggressive behaviour in captive Atlantic bottlenosed dolphins (*Tursiops truncatus*). *Zoo Biology, 2*(2), 93–103.

Pagel, C.D. (2015). *Interactive behaviour of killer whales (*Orcinus orca*) towards human divers in northern Norway* (Unpublished master's thesis). Goettingen, Germany and Lincoln, New Zealand, Georg-August-Universitaet and Lincoln University.

Pagel, C.D., Scheer, M., & Lück, M. (2017). Swim encounters with killer whales (*Orcinus orca*) off Northern Norway: Interactive behaviours directed towards human divers and snorkelers obtained from opportunistic underwater video recordings. *Journal of Ecotourism*. doi: 10.1080/14724049.2016.1273939

Pagel, C.D., Scheer, M., & Lück, M. (2018). A rejoinder to comments on 'Swim encounters with killer whales (*Orcinus orca*) off northern Norway: Interactive behaviours directed towards human divers and snorkelers obtained from opportunistic underwater video recordings'. *Journal of Ecotourism, 16,* 190–200. doi: 10.1080/14724049.2017.1375285

Perrine, D. (1998). Divers and dolphins. *Sport Diver, 6,* 41–47.

Pryor, K. (1990). Non-acoustic communication in small cetaceans: Glance, touch, position, gesture, and bubbles. In J.A. Thomas & R.A. Kastelein (Eds.), *Sensory abilities of cetaceans* (pp. 537–544). New York, NY: Plenum Press.

Ridgway, S.H. (1990). The central nervous system of the bottlenose dolphin. In S. Leatherwood & R.R. Reeves (Eds.), *The bottlenose dolphin* (pp. 69–97). San Diego, CA: Academic Press.

Samuels, A., Bejder, L., & Heinrich, S. (2000). A review of the literature pertaining to swimming with wild dolphins (Report). Silver Spring, MD: Marine Mammal Commission.

Scheer, M. (2010). Review of self-initiated behaviors of free-ranging cetaceans directed towards human swimmers and waders during open water encounters. *Interaction Studies, 11,* 442–466.

Scheer, M., de Sá Alves, L.C.P., Ritter, F., Azevedo, A.F., & Andriolo, A. (2014). Behaviors of botos and short-finned pilot whales during close encounters with humans: Management implications derived from ethograms for food-provisioned versus unhabituated cetaceans. In J.B. Samuels (Ed.), *Dolphins: Ecology, behavior and conservation strategies* (pp. 1–36). New York, NY: Nova Science Publishers.

Scheer, M., Hofmann, B., & Behr, I.P. (2004). Ethogram of selected behaviors initiated by free- ranging short-finned pilot-whales (*Globicephala macrorhynchus*) and directed to human swimmers during open water encounters. *Anthrozoös, 17*(3), 244–258.

Shackley, M. (1996). *Wildlife tourism*. London, England: International Thomson Business Press.

Shackley, M. (2001). *Flagship species: Case studies in wildlife tourism management*. Burlington, VT: The International Ecotourism Society.

Shane, S. (1990). Behaviour and ecology of the bottlenose dolphin at Sanibel Island, Florida. In S.L. Leatherwood & R.R. Reeves (Eds.), *The bottlenose dolphin* (pp. 245–265). San Diego, CA: Academic Press.

Shane, S.H. (1995). Human-pilot whale encounter: Update. *Marine Mammal Science, 11,* 115.

Shane, S.H., Tepley, L., & Costello, L. (1993). Life-threatening contact between a woman and a pilot whale captured on film. *Marine Mammal Science, 9*, 331–336.

Similä, T., Holst, J.C., & Christensen, I. (1996). Occurrence and diet of killer whales in northern Norway: Seasonal patterns relative to the distribution and abundance of Norwegian spring-spawning herring. *Canadian Journal of Fisheries and Aquatic Sciences, 53*, 769–779.

Spong, P. (1974). Mind in the Waters. In J. McIntyre (Ed.), *A Book to Celebrate the Consciousness of Whales and Dolphins*, (240 pp.) New York: Charles Scribner's Sons and San Francisco: Sierra Club Books.

Spradlin, T.R., Barre, L.M., Lewandowski, J.K., & Nitta, E.T. (2001). Too close for comfort: Concern about the growing trend in public interactions with wild marine mammals. *Marine Mammal Society Newsletter 9*(3). Retrieved from http://sanctuaries.noaa.gov/dolphinsmart/pdfs/spradlin_2001.pdf

Stenersen, J., & Similä, T. (2004). Norwegian killer whales. Henningsvaer/Norway: Tringa forlag.

Stringham, S.F. (2011). Aggressive body language of bears and wildlife viewing: A response to Geist (2011). *Human–wildlife interactions, 5*(2), 177–191. Retrieved from http://digitalcommons.usu.edu/cgi/viewcontent.cgi?article=1173&context=hwi

United Nations Environment Programme/Conservation of Migratory Species. (2017). *Recreational in-water interactions with aquatic mammals* (COP12/Doc.24.2.5). Nairobi, Kenya: Author.

Williams, R., Trites, A., & Bain, D.E. (2002). Behavioural responses of killer whales (*Orcinus orca*) to whale-watching boats: Opportunistic observations and experimental approaches. *Journal of Zoology, 256*, 255–270.

Xian, Y.-J., Wang, K.-X., Jiang, W.-H., Zheng, B.-Y., & Wang, D. (2010). Ethogram of Yangtze finless porpoise calves (*Neophocaena phocaenoides asiaeorientalis*). *Zoological Research, 31*, 523–530.

Yasmine M. Elmahdy, Mark B. Orams and Michael Lück

Towards more sustainable marine mammal tourism in New Zealand – reviewing the literature and identifying the gaps

1. Introduction

Marine mammal tourism in New Zealand has grown rapidly, with tourists utilising different viewing platforms (e.g. boat-based, land-based, air-based and in-water encounters) to view, swim with, and photograph marine mammals. New Zealand has been often held up internationally as a model country, having both the *Marine Mammals Protection Act* (MMPA; New Zealand Government, 1978) and associated *Marine Mammals Protection Regulations* (MMPR; New Zealand Government, 1992), which aim to protect and conserve marine mammals via controlling and managing all marine mammal tourism operations. Despite this framework, a wide range of studies conducted in New Zealand have shown that marine mammal tourism activities have had adverse effects on the targeted species due to alterations in behaviour, movement and reproductive success. Furthermore, various studies have demonstrated that the regulations are frequently violated by marine mammal tour operators. Consequently, the sustainability of the industry is in question. This chapter provides a review of marine mammal tourism, its management and its impacts on marine mammals within the New Zealand context.

2. Marine-mammal tourism worldwide

Marine mammal tourism comprises "tours by boat, air or from land, with some commercial aspect, to see and/or listen to marine mammals" (Jefferies, 2016, p. 137). In recent decades, tourists' desire to view and interact with marine mammals in the wild has driven strong growth in marine mammal tourism (Constantine & Bejder, 2008; Lundquist, 2014; Martinez, Orams, & Stockin, 2011; Orams, 1997). Cetaceans (whales, dolphins and porpoises) are the main target species for the marine mammal tourism industry (Dans, Crespo, & Coscarella, 2017); however, other marine mammals such as pinnipeds (seals, sea lions and walruses; Dans et al., 2017; Newsome & Rodger, 2008), manatees

(Sorice, Shafer, & Ditton, 2006) and polar bears (Lemelin & Dyck, 2008) are becoming increasingly targeted by tourism activities as well.

Interactions between visitors and marine mammals in the wild can take various forms (Lusseau, 2008). Visitors can observe whales, dolphins, otters and pinnipeds from a vessel on the water. Tourism encounters can also take place on land with whales (Hoyt, 2001, 2009), polar bears, otters and pinnipeds either by foot or utilising a variety of motorised vehicles such as quad bikes, tundra vehicles and helicopters (Lusseau, 2008).

An upsurge and spread of the nature-based marine tourism industry that revolves around free-ranging cetaceans (whales, dolphins and porpoises) (Allen, 2014) (hereafter referred to as whale-watching) has occurred over the past three decades. The first commercial whale-watching tour took place in California, USA, in 1955 and charged US$ 1 to view grey whales (*Eschrichtius robustus*) on their winter migration off San Diego (Hoyt, 2009). Thenceforward, interest and demand for whale-watching grew steadily in the US and later expanded rapidly to other regions (Hoyt, 2009). Today, whale-watching takes place throughout the world and in countries as varied as Canada, Norway, Brazil, Japan, Indonesia, South Africa, New Zealand and the Cook Islands (Orams, 2000, 2004). Different viewing platforms such as boat-based, land-based, air-based and in-water encounters are currently used to watch, photograph and swim with cetaceans (Higham & Hendry, 2008; Hoyt, 2001, 2009; Zeppel & Muloin, 2008).

The economic value of marine mammal tourism is widely recognised (Lundquist, 2014; Meissner et al., 2015). In particular, the socio-economic benefits of the whale-watching industry have sustained remarkable growth (Higham, Bejder, & Williams, 2014) with a 12 % per annum average growth rate in global whale-watch numbers reported through the 1990s along with increases in tourist expenditure of 18.6 % per annum (Garrod & Fennell, 2004; Hoyt, 2001). From approximately four million whale-watchers in only 31 countries and territories in 1991, the industry grew to around nine million participants in 87 countries and territories in 1998 (Hoyt, 2001). Whale-watching is now worth more than US$ 2.1 billion worldwide in total expenditure, 13 million participants in 119 countries, generating over 13,000 jobs provided by more than 3,300 operators (O'Connor, Campbell, Cortez, & Knowles, 2009).

Moreover, a number of authors argue that there is still plenty of room left for growth in the industry and, hence, it is expected to continue to expand in the coming years (Cisneros-Montemayor, Sumaila, Kaschner, & Pauly, 2010; Lundquist, Gemmell, & Würsig, 2012; Parsons, 2012). It has been estimated that an additional 5,700 jobs and US$400 million could potentially be generated, if maritime countries with cetacean populations that are currently without whale-watching industries, were to develop them (Cisneros-Montemayor et al., 2010; Parsons, 2012).

Despite the fact that cetaceans are the main target for marine mammal tourism (Dans et al., 2017), in recent years pinniped-based tourism has also developed and rapidly expanded (Dans et al., 2017; Kirkwood et al., 2003; Newsome & Rodger, 2008). This is a result of various behavioural attributes these animals exhibit (Dans et al., 2017; Newsome & Rodger, 2008). Generally, pinnipeds are colonial species with predictable annual patterns of attendance and residency. They display 'playful' and interactive behaviours that appeal to tourists, hence, providing a viewing spectacle. Similar to whale-watching, viewing experiences with pinnipeds range from land-based (on-shore) guided tours to boat-based tours and swim-with interactions (Dans et al., 2017; Newsome & Rodger, 2008).

According to Young (1998), around 500,000 visitors participated in 117 boat-based seal-watching tours in the UK and Ireland in 1997. While in the southern hemisphere, Kirkwood et al. (2003) reported that there were about 80 pinniped tourism sites with an economic value of approximately US$12 million per year (Kirkwood et al., 2003). Australia alone comprises around 53 operators organising visits to 23 sites and involving nearly 400,000 tourists. "Other important southern hemisphere locations include the Kaikoura Peninsula, New Zealand (~250,000 tourists per annum), Duiker Island, South Africa (~200,000 tourists per annum) and the Peninsula Valdez, Argentina (~150,000 tourists per annum)" (Newsome & Rodger, 2008, p. 182). Pinniped-based tourism activities also take place in Europe, North America and the Galapagos Islands (Newsome & Rodger, 2008).

3. Marine mammal tourism in New Zealand

New Zealand, a South Pacific country with a diverse and rich fauna of marine mammals and which is home to nearly half of the world's species of cetaceans (Department of Conservation, 2020a), has followed the global trend of growth in tourism based on these animals (Meissner et al., 2015; O'Connor et al., 2009). Marine mammal tourism in New Zealand is currently considered a significant aspect of the wider tourism offering in the country (Orams, 2004). "Since the first commercial operation began at Kaikoura in 1987 with a single six-meter vessel taking commercial tours to watch sperm whales (*Physeter macrocephalus*)", the marine mammal tourism industry has witnessed a significant increase in the number of tourists and tour operators (Constantine, 1999, p. 6).

The number of domestic and international tourists participating in whale-watching activities in New Zealand has risen from 230,000 in 1998 to approximately 550,000 tourists (9 % per annum average growth) in 2008 resulting in more than US$ 80 million in expenditure (O'Connor et al., 2009). According to Kirkwood et al. (2003), pinniped-based tourism in New Zealand earned over US$1 million and the number of tourists participating in pinniped-focused tourism activities has reached around 250,000 tourists in the Kaikoura peninsula alone. Fur seal-based tourism activities at Kaikoura range from fur seal (or dolphin) watch and swim-with boats (78 tours per week), boat-based fur seal swim-with tours (119 tours per week), land-based fur seal swim-with tours (35 tours per week) and land-based fur seal watching (21 tours per week) (Newsome & Rodger, 2008). Permits issued to watch and/or swim-with-dolphins, whales or fur seals increased from 90 in 2005 to 112 in 2011 (Meissner et al., 2015), and today commercial activities take place in over 10 locations in both North and South Islands (Cross, 2013).

Marine mammal species targeted by tourism in New Zealand (some of which are nationally endangered) comprise five species of dolphins, two species of whales and two species of pinnipeds (Cater & Cater, 2007; Constantine, 1999). These include dusky dolphins (*Lagenorhynchus obscurus*), common dolphins (*Delphinus delphis*), killer whales (*Orcinus orca*), as well as bottlenose dolphins (*Tursiops truncatus*), the endemic Hector's dolphins (*Cephalorhynchus hectori hectori*), sperm whales (*Physeter macrocephalus*), Bryde's whales (*Balaenoptera edeni*), New Zealand fur seals (*Arctocephalus forsteri*), and New Zealand sea lions (*Phocarctos hookeri*) (Constantine, 1999; Meissner et al., 2015). In addition, a variety of other cetanceans can occasionally be encountered, such as humpback whales (*Megaptera novaeangliae*), southern right whales (*Eubalaena australis*), southern right whale dolphins (*Lissodelphis peronii*), false killer whales (*Pseudorca crassidens*), blue whales (*Balaenoptera musculus*), and others, although these encounters are infrequent (Lück, 2009).

Examples of marine mammal tourism activities range from viewing cetaceans such as dusky dolphins and sperm whales in Kaikoura (Lundquist et al., 2012), and common dolphins in the Bay of Plenty (Meissner et al., 2015) or Mercury Bay (Neumann & Orams, 2006) to viewing and swimming with bottlenose dolphins in the Bay of Islands (Constantine, 2001; Constantine, Brunton, & Dennis, 2004) and Hector's dolphins in Akaroa Harbour (Martinez et al., 2011). Activities focusing on pinnipeds include swimming with New Zealand fur seals (the most commonly found pinniped in New Zealand) and viewing them from land, air or from tour vessels and kayaks (Boren, Gemmell, & Barton, 2002) in locations such as the Bay of Plenty, North Island (Cowling,

Kirkwood, Boren, & Scarpaci, 2014; Cowling, Kirkwood, Boren, Sutherland, & Scarpaci, 2015) and Kaikoura, South Island (Boren et al., 2002).

It is noteworthy that, in addition to the significant growth in commercial marine mammal tourism activities in New Zealand (Orams, 2004), marine and coastal areas are also becoming increasingly popular and easily accessible settings for a wide array of recreational activities (Hansen, 2016). According to Orams' (1999) Spectrum of Marine Recreation Opportunities (SMARO) model, the diverse range of marine recreational activities can be classified according to the experiences they provide, the type of environment in which they take place and their distance from shore. At one end of the spectrum, easily accessible near-shore activities (e.g. swimming, sunbathing) take place, while at the other end are those activities which occur far from shore, in more isolated and remote areas (e.g. offshore sailing, remote sea-kayaking). In the middle ground lie a range of experiences, environmental settings and recreational activities (e.g. sailing, snorkelling, and surfing).

In New Zealand, participation in marine recreational activities is on the rise and remote marine areas are becoming progressively more accessible. As reported by Orams (2004), there are now large numbers of private (non-commercial) recreational vessels operated by New Zealanders who seek to view and interact with marine mammals in their natural environment. In more recent times, there have also been increasing reports of New Zealanders incidentally encountering and interacting with marine mammals (e.g. whales, dolphins or seals) in the wild while out enjoying a variety of coastal activities such as kayaking, stand-up paddle boarding and surfing (Orams, 2016; Fandel, Bearzi, & Cook, 2015).

According to the *Auckland Sport and Recreation Strategic Action Plan 2014-2024* (Auckland Council, 2014), growth in participation in outdoor (marine) recreation in New Zealand can be attributed to a variety of factors such as population growth and new technologies. New Zealand's population is growing and becoming more diverse (Auckland Council, 2014). For example, in 2013, Auckland's population reached 1.4 million people, accounting for 33.4 % of New Zealand's population and it is expected to reach around 2.5 million by 2040 (Auckland Council, 2014). Growth and the changing composition of the population are expected to continue changing outdoor recreation participation patterns while also creating more pressure on existing natural resources and facilities (Auckland Council, 2014).

New technologies are another important driver behind the increasing participation in outdoor recreation (Buckley, 2000). According to Buckley (2000), the development and introduction of new high-tech recreational equipment (e.g. GPS, marine sonars, chartplotters) and outdoor clothing (e.g. GORE-TEX® and POLARTEC® fabric technology) are main drivers behind the immense growth in outdoor recreation. As reported by Shultis (2012, 2015), the embedded technology within the different types of recreational equipment/clothing is empowering for outdoor (marine) recreationists, because it provides them with increased safety, access to natural areas and comfort with warmth, dryness and lightweight materials (Shultis, 2012, 2015). This may also allow for year-round participation in marine recreational activities in various locations (including remote areas), consequently leading to more frequent encounters with marine mammals in the wild.

Such patterns and rates of growth of marine mammal tourism and recreation in New Zealand and elsewhere have raised concerns over impacts on both the animals and tourists (Constantine, 1999; Garrod & Fennell, 2004; New et al., 2015; Orams, 2004, 2005). In many locations, the rapid development of marine mammal tourism (especially whale-watching) has outpaced management, resulting in concerns over the long-term sustainability of the industry (Garrod & Fennell, 2004; New et al., 2015). The desire to mitigate the potential for disturbance has led to the introduction and development of different management policies, guidelines and regulations (Garrod & Fennell, 2004; Lück & Higham, 2008; New et al., 2015; Parsons, 2012), which have been introduced in more than 100 countries (as claimed by New et al., 2015).

The following presents a summary of the different approaches used to manage the marine mammal tourism industry worldwide. However, it is deemed important to define the concept of *sustainable management* prior to reviewing these management strategies. According to New Zealand's *Resource Management Act* (RMA; New Zealand Government, 1991), which has the purpose of promoting the sustainable management of environmental resources, sustainable management is defined as:

> Managing the use, development, and protection of natural and physical resources in a way, or at a rate, which enables people and communities to provide for their social, economic, and cultural well-being and for their health and safety while—
> *(a) sustaining the potential of natural and physical resources (excluding minerals) to meet the reasonably foreseeable needs of future generations; and*
> *(b) safeguarding the life-supporting capacity of air, water, soil, and ecosystems; and*
> *(c) avoiding, remedying, or mitigating any adverse effects of activities on the environment. (p. 62)*

It is argued that for sustainable development to be achieved, it is important to harmonise three key elements: social inclusion, economic growth and environmental protection. These elements are interconnected and all are essential for the welfare of both individuals and societies (United Nations, 2018).

4. Marine mammal tourism management approaches

Management strategies used to control and manage marine mammal tourism vary from one country to another (Orams, 2000). Management approaches range from inadequate or completely lacking (no control whatsoever) at one extreme, to complex 'command and control' regulations at the other (Allen, Smith, Waple, & Harcourt, 2007; Constantine & Bejder, 2008; Orams, 1999). In between these two extremes lies a range of semi-formal, voluntary measures that include 'codes of conduct' or 'guidelines' (Allen et al., 2007; Garrod & Fennell, 2004).

4.1 Unregulated marine mammal tourism

The growth of the marine mammal tourism industry is now driven by the expansion of destinations in developing nations (Higham, Bejder, Allen, Corkeron, & Lusseau, 2016; Hoyt & Parsons, 2014), where regulatory methods to manage and control the industry are either very limited or non-existent (e.g. Beasley, Bejder, & Marsh, 2014; Mustika, Birtles, Everingham, & Marsh, 2013; Mustika, Birtles, Welters, & Marsh, 2012). For instance, the critically endangered population of Irrawaddy dolphins (*Orcaella brevirostris*) inhabiting the Mekong River in Cambodia and Laos are being chased and harassed by an increasing number of tour vessels (Beasley, Bejder, & Marsh, 2010; Beasley et al., 2014). According to Beasley et al. (2014), the governments of these two countries have never attempted to facilitate management to reduce boat disturbance on the targeted dolphins.

Unregulated whale-watching has also been documented at Lovina in north Bali, Indonesia, where around 179 *Jukungs* (traditional tour boats) conduct tours targeting mainly spinner dolphins (*Stenella longirostris*) (Mustika, 2011; Mustika et al., 2012, 2013). Studies by Mustika et al. (2012, 2013) have shown that the number of boats in an encounter often exceeds the number of dolphins (dolphin to boat = 0.8:1). Moreover, the boatmen at Lovina tend to speed, get too close or chase the dolphins (Mustika et al., 2013, 2012), risking injuring them (Mustika, Birtles, Everingham, & Marsh, 2015). Similarly, unmanaged whale-watching activities have been reported at Chilika Lagoon, India, where about 900 converted fishing vessels target an isolated and declining (<150 animals) population of Irrawaddy dolphins (Beasley et al., 2014; D'Lima, 2015; Mustika et al., 2017).

The findings of the above-mentioned studies provide examples of unmanaged and unregulated marine mammal tourism in a number of developing countries. The lack of management of the industry in these countries has been associated with factors such as inadequate governance, conflicting policy goals, limited in-country capacity, and insufficient accountability (Beasley et al., 2014). The concern over the rapid expansion of unregulated marine mammal tourism activities (especially whale-watching) and the adverse effects that could occur to cetacean populations, particularly in developing countries, have led researchers to recommend adopting the precautionary principle to management to tackle the problem of unsustainable whale-watching in these countries (Beasley et al., 2010, 2014). Moreover, it has prompted the Whale-watching Subcommittee of the International Whaling Commission (IWC) to adopt this subject as an annual agenda item (Hoyt & Parsons, 2014).

4.2 *Semi-regulated marine mammal tourism*

Concerns over the negative impacts of tourism on targeted species has prompted many other countries to informally intervene via developing and implementing a set of self-regulatory environmental management tools known as *codes of conduct* (Garrod & Fennell, 2004; Gjerdalen & Williams, 2000). Codes of conduct are developed by either governments, local communities, regional, national and/or local tourism industry groups, tour operators, tour guide associations, non-governmental organisations (NGOs) or a combination of any of these stakeholders (Cater, Garrod, & Low, 2015).

Such codes can be described as written records defining specific guidelines of behaviour for distinct groups of stakeholders (Garrod & Fennell, 2004; Gjerdalen & Williams, 2000) including tourists or visitors, tour operators and tour guides (Cater et al., 2015). They are voluntary measures which are enforced mainly by peer pressure or ethical obligation (Garrod & Fennell, 2004). According to Garrod and Fennell (2004), two-thirds of international whale-watching guidelines and codes of conduct are entirely voluntary, while only one-third are regulatory (non-voluntary) measures with legal requirements (Garrod & Fennell, 2004; Parsons, 2012). Hence, voluntary codes of conduct are by far the most commonly used management strategy (Heenehan, Van Parijs, Bejder, Tyne, & Johnston, 2017).

Gjerdalen and Williams (2000) state that codes of conduct such as ones developed for whale-watching activities can have a number of benefits. For example, they can be effective in promoting stewardship, supporting local tourism and helping different stakeholders carry out their activities in a righteous and ethically-responsible manner. Additionally, they may also be utilised to manage whale-watching while more formal regulations are being formulated and introduced (Cater et al., 2015; Garrod & Fennell, 2004). However, it is contended that these voluntary codes also have their shortcomings.

Garrod and Fennell (2004) argue that, even though the majority of codes of conduct have minimum approach distance requirements, most do not succeed in mitigating invasive activities by, for instance, inhibiting feeding or touching marine mammals (Garrod & Fennell, 2004). Furthermore, since these codes only suggest how users should behave, not whether they actually conform, some individuals may deliberately violate the codes for reasons such as ideological protest, monetary gain, revenge or just for fun. Other factors associated with the violation or non-compliance of codes of conduct may include insufficient awareness of expected behavioural norms, inadequate awareness of the negative environmental impacts associated with the violations, improper environmental behaviour cues received from other individuals or the codes being perceived as unimportant or impractical (Gjerdalen & Williams, 2000).

A number of studies have highlighted the weakness of voluntary guidelines or codes of conduct in managing the industry (Allen et al., 2007; Duprey, Weir, & Würsig, 2008; Kessler & Harcourt, 2010). In many cases, tour operators are found to be pressured to provide tourists with an ideal experience, which often leads to a tendency to ignore the guidelines (Kessler & Harcourt, 2010). For example, in Tonga, Kessler and Harcourt (2010) reported that tour operators conducting swim-with whales activities focusing on an endangered subpopulation of humpback whales (*Megaptera novaeangliae*) followed the voluntary guidelines to different extents. Moreover, swimming with calves was also very common, which has raised concerns over the potential negative impact of these activities on the breeding success of the targeted endangered subpopulation of whales. Similarly, Allen et al. (2007) found that in Port Stephens, Australia, some tour operators conducting boat-based tours to watch and interact with Indo-Pacific bottlenose dolphins (*Tursiops aduncus*) did not comply with all the conditions of the adopted code of conduct. Non-compliance activities included a lack of distinction between groups containing calves and those that did not and repeated exposure of dolphins to various dolphin-watching commercial tour operators and other vessels.

Researchers argue that voluntary codes are often of minimal value without enforcement and recommend that governments (e.g. Government of Tonga) adopt strong regulatory frameworks to enforce appropriate behaviour around marine mammals to ensure the sustainability of the industry (Allen et al., 2007; Kessler & Harcourt, 2010).

4.3 Regulated marine mammal tourism

Another conventional approach to managing the industry is via adopting what is known as the 'command and control' regulatory management approach (Constantine & Bejder, 2008; Garrod & Fennell, 2004; Orams, 1999). In some countries (e.g. New Zealand – Hoyt & Parsons, 2014; Lück, 2008b; Orams, 2004, 2005), such regulatory management practices are a commonly used method of governing tourist activities in marine environments (Lück, 2008b; Orams, 1999). These management practices "refer to government measures (legislation) with which individuals or corporate entities must comply" (Davis & Gartside, 2001, p. 224). Regulations frequently utilised to manage visitors or tourists include restrictions on types of visitor activities allowed, limits on locations and times, limits on numbers of visitors, temporal and spatial zoning, restrictions on the types of equipment allowed, speed limits, noise levels, and requirements for permits and/or licenses (Orams, 1999).

These regulatory management systems usually adopt a permit system and/or punitive policies which are used if the regulations are violated (Lück, 2008b). Enforcement of these regulations often falls to police, rangers or other officers belonging to a management authority (Orams, 1999). Moreover, an increasingly harsh punishment system is most common in case of transgressions; for instance, at a first-time offence, a warning is given, at a second offence, punishment is in the form of a ban from the area, and ultimately fines and/or imprisonment for major cases of misconduct which involve for instance harassing or killing protected wildlife (Orams, 1999).

Regulatory practices have been judged as an effective management approach to preserve marine mammals and the environment, to ensure tourist safety and to mitigate conflicts among different stakeholders (Orams, 1999). However, it is argued that there are still concerns that such regulations have either insufficient or insignificant effects on the mitigation of negative impacts of tourism activities on marine mammals (Forestell, 2008). Possible reasons for that may be that either an impact of concern is not properly addressed by the regulations or the regulations are not being correctly followed (Forestell, 2008). According to New et al. (2015), mandatory guidelines are hard to enforce, and therefore non-compliance by both recreational and commercial boats is common.

New Zealand provides an example of a country that adopts this formal regulatory management approach (Hoyt & Parsons, 2014; Lück, 2008b; Orams, 2004,

2005). At an early stage, New Zealand regulated its marine mammal tourism industry with a set of operating regulations and a permit scheme (Gjerdalen & Williams, 2000; Hoyt & Parsons, 2014) based on precautionary management and scientific studies (Hoyt & Parsons, 2014). It is, therefore, deemed worthwhile to review the marine mammal tourism management regime utilised in New Zealand as it is regarded as one of the strongest in the world (Orams, 2004, 2005).

5. Marine mammal tourism management in New Zealand

In New Zealand, marine mammals are granted full protection under the *Marine Mammals Protection Act* (MMPA; New Zealand Government, 1978). Marine mammal tourism (and recreation) is managed under the *Marine Mammals Protection Regulations* (MMPR; New Zealand Government, 1992). The purpose of these regulations is to protect, conserve and manage marine mammals. They also aim to regulate human (either commercial operators or other persons) interactions and behaviour with marine mammals.

5.1 *Marine Mammals Protection Regulations (MMPR)*

In order to minimise the increasing threats posed by marine mammal tourism and recreational activities such as harassment, boat strike, noise pollution, displacement and separation of mothers and their young, the *Marine Mammals Protection Regulations* (MMPR; New Zealand Government, 1992) prescribe appropriate behaviour around marine mammals. Generally, individuals or groups must not harass, disturb or make any loud noises near marine mammals. They must not feed the animals or throw any rubbish near them, and contact must be terminated if the targeted species exhibit any signs of disturbance. Furthermore, the regulations also dictate proper behaviour appropriate for different settings: at sea, in the air and on shore (New Zealand Government, 1992).

5.1.1 *Behaviour around marine mammals – at sea*

As illustrated in Figure 1, tour vessels and recreational vessels should not travel faster than at idle or 'no wake' speed within 300 metres of marine mammals and their number should not exceed three vessels. A vessel may gradually increase the speed to outdistance dolphins and should not exceed 10 knots within 300 meters of any dolphin. The vessels should approach dolphins or whales from behind and the side, and should not encircle them, block their path or cut through any group. When viewing whales, a tour vessel should keep at least

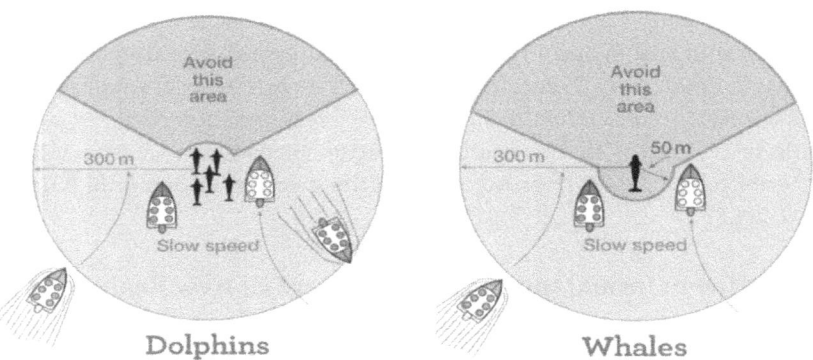

Figure 1. Rules around marine mammals. Reprinted from Department of Conservation (2017b)

50 metres from the whales or 200 metres from any whale mother accompanied by a calf or calves. Swimming with whales or orcas is not permitted; however, swimming with dolphins or pinnipeds is allowed but not with juvenile dolphins or a pod of dolphins that includes juvenile dolphins (New Zealand Government, 1992).

Additionally, according to the Department of Conservation's website, vessels need to stay at least 20 metres away from the water's edge where seals may be present, and swimmers need to stay at least five metres away from the water's edge (Department of Conservation, 2020b). However, upon reviewing the MMPR in general and the 'Special conditions applying to dolphins or seals' section in particular, these two latter seal-related rules were not found anywhere in the document (New Zealand Government, 1992, p. 14). Therefore, indicating a disparity between information provided by the Department of Conservation on their official website and the official MMPR document.

5.1.2 Behaviour around marine mammals – in the air

In the air, an aircraft should maintain a horizontal distance greater than 150 metres, and an altitude of at least 600 metres (2,000 feet) when flying near any marine mammal. Aircrafts should also refrain from flying or imposing a shadow directly over marine mammals either on shore or at sea (New Zealand Government, 1992). In regard to drones or UAVs (unmanned aerial vehicle) rules around marine mammals, according to the Department of Conservation's website, these should have the same restrictions as aircraft (Department of Conservation, 2020b). However, it is noteworthy to point out that this is not stated in the MMPR (New Zealand Government, 1992).

5.1.3 Behaviour around marine mammals – on shore

As for viewings and interactions which occur on shore, individuals or groups should stay at least 20 metres away from fur seals and sea lions and if accompanied by dogs, the dogs should be kept on a leash at all times. Visitors should also avoid coming between fur seals and sea lions and the sea and should never attempt to touch them (Department of Conservation, 2020b). Again, these rules are not mentioned anywhere in the MMPR (New Zealand Government, 1992), indicating that they are only voluntary guidelines as opposed to enforceable regulations. If a vehicle is utilised and if practicable, visitors should not drive within 50 metres of a marine mammal (New Zealand Government, 1992).

The Department of Conservation (DoC) is responsible for administering these laws and regulations (New Zealand Government, 1978, 1992; Orams, 2004, 2005), and its main mechanism for doing so is through issuing marine mammal tourism permits (Orams, 2004, 2005).

5.1.4 Marine mammal tourism permits

Any commercial enterprise wishing to promote and provide marine mammal tourism activities such as observing or swimming with marine mammals is required to have a permit (Lück, 2008b; Orams, 2004, 2005; Sun & Lück, 2015). Permits can have a range of conditions attached to them and criteria under which they are issued.

The main 'criteria' that need to be fulfilled before a permit is issued are as follows: all permits require a tour operator to have no significant negative impact on the targeted species and be for their conservation, protection and management. An operator and any of the operator's staff, who may come in contact with marine mammals, should have adequate experience with marine mammals and satisfactory knowledge of the local area, weather conditions and the sea. An operator should ensure that a commercial operation has sufficient educational value to participants or to the public. The operator and staff members who may come in contact with marine mammals should not have convictions for offences which involve maltreatment of animals. If all permit issuance criteria are met, then a permit is issued based on the specific plan of operation which is required as part of the lodged application (Constantine, 1999; New Zealand Government, 1992).

In regard to the 'conditions' related to marine mammal tourism permits, these may include restrictions on locations where tourism activities occur, species targeted, status of targeted species (e.g. vessels not allowed to approach mothers with calves), time spent with animals, types of vessels utilised, types of vessel propulsion, vessels speed, minimum approach distances for different species, minimum depths, maximum number of tour vessels within a certain range, maximum number of swimmers, and maximum number of tours (Orams, 2004).

The flexibility associated with the permitting process and permit-related conditions has been judged advantageous in allowing DoC to adapt management regimes to suit specific species, locations or vessel types (Orams, 2004). Yet this flexible system has been also criticised as problematic due to the absence of consistency around the country. This usually results in a lack of awareness of the restrictions and conditions among non-permitted tour operators or private recreational vessels which often leads to conflicts among different stakeholders such as permitted and non-permitted operators (Orams, 2004).

Even though non-compliance with the regulations may be due to a lack of awareness among, for instance, recreational or non-permitted (scenic) tour vessels interacting with marine mammals (Orams, 2004), still they are bound by the *Marine Mammals Protection Regulations* (MMPR). Consequently, any failure to comply with the regulations by either commercial operators or other persons such as private recreationists may result in fines reaching NZ$ 250,000 or penalties of up to two years imprisonment (Department of Conservation, 2020b; New Zealand Government, 1992).

6. Impacts of marine mammal tourism

Watching marine mammals in their natural environment is regarded as a positive way to provide visitors with conservation, educational and psychological benefits (Zeppel & Muloin, 2008, 2014), in addition to the economic benefits in the form of tourism revenues (Orams, 1999; Zeppel & Muloin, 2008). This sector has the potential to produce meaningful contributions to local and national economies. It can have a positive economic impact on local communities by providing a range of employment opportunities; tourists spend large amounts of money on different products and services, in addition to the wider economic value connected to the conservation of the different nature tourism destinations (Lück, 2008a; Zeppel & Muloin, 2008).

Furthermore, interactions between tourists and marine mammals in the wild can provide conservation and educational benefits by conveying useful information and knowledge about marine wildlife through interpretation programmes (Lück, 2008a, 2015; Zeppel & Muloin, 2008, 2014). These programmes have the capacity to change visitors' pro-environmental attitudes and beliefs if organised and carried out efficiently (Lück, 2003, 2008a; Orams, 1994, 1997; Zeppel & Muloin, 2008, 2014). Additionally, such interactions provide visitors with various psychological benefits such as excitement, novelty and an enhanced personal well-being (Johnson & McInnis, 2014; Lück, 2008a; Zeppel & Muloin, 2008, 2014). Marine mammal tourism occasionally provides support for natural and/or social science research (e.g. in New Zealand) as well via allowing researchers on board tour vessels to conduct research and research-related activities (Orams, 2004, 2005).

Marine mammal tourism is viewed by many as a legitimate form of nature-based tourism (Orams, 2000, 2005). It is often perceived as an environmentally sound and conservation friendly method to utilise marine wildlife and the environment (Currey et al., 2009; Garrod & Fennell, 2004; Martinez et al., 2011). Moreover, the industry is frequently promoted as a non-consumptive activity (Garrod & Fennell, 2004; Higham et al., 2016), a viable alternative to lethal whaling (Garrod & Fennell, 2004; International Fund for Animal Welfare, 1995; New et al., 2015; Orams, 2004, 2005) or to watching wild animals in captivity (Hughes, 2001).

However, the public appetite for watching marine mammals in the wild has become voracious (Higham et al., 2014) and a variety of studies have shown that marine mammal tourism activities are far from benign (Higham et al., 2016; Orams, Forestell, & Spring, 2014). The growth of the industry has raised various concerns over its potential negative impact on marine wildlife and their environment (Acevedo-Gutierrez, Acevedo, Belonovich, & Boren, 2010; Garrod & Fennell, 2004). "Hence, the paradox: more tourism provides benefits to wildlife yet it also poses increased dangers" (Acevedo-Gutierrez et al., 2010, p. 39).

Similarly, concerns over the negative impact of tourism operations on marine mammals have been raised also in New Zealand where the marine mammal tourism industry is highly regulated, and which is considered a 'model' country in conservation and protection of marine mammals. Such concerns have consequently led researchers to conduct and complete a number of studies assessing the impacts of tourism on different species in various locations nationwide. The following provides a general overview of the negative impacts of tourism on marine mammals followed by an overview focusing on impact studies in New Zealand.

7. Negative impacts of marine mammal tourism on marine mammals

Marine mammal tourism has been lately judged as a typical capitalist mode of production which prioritises profit over ecological conservation (Higham et al., 2016; Higham & Neves, 2015). In addition, the same authors report that in recent years, the industry has often been associated with negative environmental impacts, exploitation of natural resources and the proliferation of 'unequal development'. A variety of studies have shown that marine tourism activities have adverse short-term and long-term effects on marine mammals (Lundquist, 2014; Lusseau, 2006; New et al., 2015; Orams, 2004).

Tourists' interactions with marine mammals affect their behaviour (Currey et al., 2009; Martinez et al., 2011); their movement becomes more unpredictable and their socialising and resting behaviour is disturbed (Currey et al., 2009). Short-term negative impacts upon cetaceans associated with boat disturbance and interacting with tourists have been examined in various studies (Lusseau, 2006; Lusseau & Higham, 2004). During these interactions, cetaceans such as dolphins often show signs of active avoidance; both horizontally and vertically. Avoidance behaviour ranges from changes in acoustic communication, increases in swimming speed (Lusseau & Higham, 2004), increases in dive intervals, increases in erratic movements and displacement (Lusseau, 2006). Long-term negative impacts of marine mammal tourism activities on targeted species have been also documented in the literature (Lundquist et al., 2012; Lusseau, Slooten, & Currey, 2006). These include alteration in residency patterns or area avoidance (Lundquist et al., 2012), a reduction in reproductive success and a decline in population-level (Bejder et al., 2006; Lundquist et al., 2012).

Cetaceans' avoidance behaviour existence and frequency is not only due to the presence and quantity of boats but is also a result of their unpredictable movement (Lusseau & Higham, 2004). In other words, the more boats are manoeuvred unpredictably, the more the animals try to escape and avoid them (Higham & Lück, 2008). Many marine mammal species associate the erratic navigation of boats as a threat to members of the group. Therefore, they demonstrate anti-predatory techniques when boats approach and attempt to block their way or hinder their movement (Higham & Lück, 2008). Studies have also shown that marine mammals' (especially cetaceans) exposure to noise pollution from tourism platforms, particularly tour vessels, may negatively affect different aspects of their life such as their foraging abilities and reproductive success (Lusseau, 2008). Moreover, Lusseau (2008) found that cetaceans' exposure to noise from tour vessels results in temporary physical damage (temporary threshold shift or TTS) and it is believed that during the high tourism season cetaceans are in a constant TTS state which may lead to a permanent threshold shift (PTS).

Furthermore, disruption caused by tourism activities has been shown to disturb the resting patterns of marine mammals including dusky dolphins (Lundquist et al., 2012), bottlenose dolphins (Currey et al., 2009; Lusseau, 2006), spinner dolphins (Shawky & Afifi, 2008; Timmel, Courbis, Sargeant-Green, & Markowitz, 2008) and fur seals (Cowling et al., 2014). Resting is vital for animals (Constantine et al., 2004) and for the majority it occurs during the daytime, which is when human disturbance from tourism occurs (Shawky & Afifi, 2008). Any interruption or cutback in resting may cause significant negative responses, such as reduced energy reserves, increased energetic costs and increased physiological stress (Constantine et al., 2004; Lusseau, 2006). In the long term, these responses can have detrimental ramifications for reproductive success and survival; both at the individual and population level (Lusseau, 2006).

8. Impact studies in New Zealand

8.1 *Bottlenose dolphins*

Examples from New Zealand include evidence that when exposed to boat disturbance and interactions with tourists, bottlenose dolphins in Doubtful Sound, Fiordland, resting patterns were significantly disrupted (Lusseau et al., 2006). During these interactions, the dolphins showed signs of active avoidance; both vertically and horizontally. Avoidance behaviour ranged from increases in swimming speed, changes in acoustic communication (Lusseau & Higham, 2004), increases in erratic movements, increases in dive intervals and displacement (Lusseau, 2006). Moreover, tour boats were found to violate the regulations (MMPR) by being intrusive via impeding the movement of the dolphins which caused an increase in the dolphins' avoidance behaviour (Lusseau, 2006). Consequently, Lusseau (2006) concluded that "it is recommended to respect the MMPR guidelines because they help minimize the impact of interactions" (p. 813).

In Milford Sound, Fiordland, bottlenose dolphins displayed similar behavioural responses to tour vessels (Lusseau et al., 2006). The dolphins have also displayed alterations in residency patterns in response to tour vessels' presence, indicating long-term negative impacts on the populations (Lusseau, 2006; Lusseau et al., 2006). Furthermore, Lusseau et al. (2006) found that both permitted and non-permitted operators violated the regulations (MMPR), the former by being intrusive and the latter by significantly interacting with the dolphins while not having a permit to do so. As a consequence, Lusseau et al. (2006) commented upon the current marine mammal management regime stating that:

> Although the New Zealand government advocates the precautionary principle in the management of natural resources this is not being applied in the context of the development of boat-based tourism activities in Fiordland. We are urging the New Zealand government to take actions to protect the small and isolated populations of bottlenose dolphins in Fiordland. (p. 177)

In the Bay of Islands, Constantine et al. (2004) found that an isolated population of bottlenose dolphins (400–500 dolphins) rested less as tour vessel numbers increased. The dolphins were also found to spend more time milling and resting less in the company of permitted dolphin-watching vessels in comparison to non-permitted vessels. Moreover, a further reduction in the dolphins' resting behaviour has been documented as a result of an increase in the number of permitted trips per week and an alteration in their departure times. Constantine et al. (2004) therefore concluded that "the implementation of the Marine Mammals Protection Act (1978) in the Bay of Islands is not affording the dolphins' sufficient protection from disturbance" (p. 305) and that "there is a need to produce effective management reform within a short time-frame" (p. 305).

A more recent study conducted by Peters and Stockin (2016) confirmed that the current high level of vessel interactions (permitted, non-permitted and private) with bottlenose dolphins in the Bay of Islands are unsustainable and have adverse impacts on the nationally endangered group of dolphins. In the presence of vessels, bottlenose dolphins spent more time socialising, diving and milling and less time resting and foraging (biologically important behaviours). The findings indicate sensitisation to tour vessel interactions through increased socialising with disruption to critically important behaviours. Non-compliance to the regulations (MMPR) by all types of vessels utilising the Bay of Islands was also documented. Furthermore, the study found that there is a local decline (65.5 % decline since 1999 and 39.6 % decline since 2005) in the numbers of bottlenose dolphins in the Bay of Islands and that there is a worrying increase in calf mortality. According to Peters and Stockin (2016), "current mitigation efforts have not been successful" (p. 15) and "clear and/or easy to follow regulations are more likely to be respected" (p. 15).

In an attempt to protect bottlenose dolphins in the Bay of Islands, the Department of Conservation put in place new permit conditions for permitted commercial tour operators (New Protection for bottlenose dolphin, 2019).

According to a media release on the Department of Conservation (DoC)'s website, these new permits came into effect on 1 July 2019 and reduce viewing and interaction time, limit the locations for these activities and prohibit swimming with bottlenose dolphins (New Protection for bottlenose dolphin, 2019). While permitted marine mammal tour operators are issued permits with conditions attached to them that they need to abide by, non-permitted commercial tour operators, private recreational boaties and other user groups' on-water behaviour is not restricted by such permit conditions. Therefore, the proposal to create a marine mammal sanctuary for the Bay of Islands which is currently being investigated by DoC (New Protection for bottlenose dolphin, 2019) would be instrumental as it would create a permanent refuge for the dolphins and ensure that all user groups (including both commercial and recreational vessels) are on an equal footing.

It is noteworthy to mention that even though the ban on swim-with-dolphins activities in the Bay of Islands appears to be well received by both DoC staff and researchers, this is not necessarily the case for some permitted commercial tour operators who displayed dissatisfaction with this decision and questioned whether it was based on sound advice (Piper, 2019). Thus, pointing to a possible conflict of interest between the different stakeholder groups involved in the marine mammal tourism industry in the Bay of Islands.

8.2 *Common dolphins*

Neumann and Orams (2006) investigated the effects of the approach of tour vessels and swimmers in the water on short-beaked common dolphins in Mercury Bay, North Island. They found that the dolphins initially approached the vessels, and then showed neutral behaviour followed by avoidance behaviour. Smaller dolphin pods were found to avoid vessels more frequently and sooner than larger pods. Furthermore, when approached by tour vessels, the dolphins altered behaviour to travelling instead of feeding (a critical behaviour). On the other hand, the authors found that the dolphins showed no signs of active avoidance of swimmers during swim-with encounters. They indicate that this may be attributed to tour operators' compliance with the regulations (MMPR) by placing swimmers around the tour vessel instead of in the dolphins' path.

For oceanic common dolphins in the Bay of Plenty, North Island, Meissner et al. (2015) reported that the dolphins' foraging behaviour was significantly changed by tour vessel interactions, potentially disrupting critical behaviour such as feeding. Disruption to this biologically important behaviour can lead to a reduction in energy intake which can have long-term negative impacts.

Furthermore, non-compliance to the MMPR by tour operators has been observed through an increase in the speed or number of vessels, manoeuvring a vessel through a pod of dolphins, exceeding of time restrictions and transgression of restrictions on swimming with calves (Meissner et al., 2015). Similar to other researchers, Meissner et al. (2015) voiced their concern by stating that since "non-compliance to the regulations (permit conditions and New Zealand MMPR) was recorded, appropriate conservation management is recommended" (p. 17).

8.3 *Hector's dolphins*

Bejder, Dawson and Harraway (1999) found that in Porpoise Bay, Hector's dolphins demonstrated an initial attraction to tour vessels lasting for up to 50 minutes. However, when an encounter lasted for more than 70 minutes, the dolphins either actively avoided the tour vessel or became equivocal towards it. Such extended encounters can potentially disrupt critical behaviours such as feeding and resting. Ultimately, they can lead to the dolphins' displacement from the bay. Moreover, the dolphins were found to form tighter pods when vessels were present, a behaviour often observed in dolphins when they feel threatened, surprised or in danger (Bejder et al., 1999).

Hector's dolphins in Akaroa Harbour, Banks Peninsula (South Island) reside in well-defined areas which make them a reliable and easy to detect target for tour operators. These endemic and endangered dolphins were found to be very receptive to contact with vessels and swimmers during swim-with tourism activities, indicating no negative consequences (Martinez et al., 2011). However, Martinez et al. (2011) confirm that even though Hector's dolphins have become more tolerant to the swimmers' presence in the water, some level of sensitisation was still apparent during the high tourism season. Hence, these dolphins are still not habituated.

Furthermore, the same authors reported that such encounters have a great potential for altering dolphins' behaviour due to the prolonged time tourists and vessels spend with dolphins (up to 18 daily swim-with dolphin tours and 14 dolphin-watching tours), especially during the high tourism season. They also stated that these tourism activities can potentially have long-term negative impacts on the dolphins by reducing time spent resting or foraging (critical behaviours). Therefore, they recommended that "the moratorium on the number of swim permits remains in place. In addition, a reduction in the level of exposure of this population of Hector's dolphins to tourism activities should be considered" (p. 99).

8.4 Dusky dolphins

Barr and Slooten (1999) found that Dusky dolphins at Kaikoura, South Island, formed tighter pods during the presence of tour vessels. They were also found to display an increase in aerial activity at a time when they are normally resting. An increase in aerial behaviour in the presence of vessels may be due to disturbance and stress or an attempt to improve visual and acoustic communication among individuals when exposed to noise pollution from vessels (Barr & Slooten, 1999). The study also reported that both commercial and private vessels violated the regulations (MMPR) by driving fast within 300 meters of a dolphin pod, blocking the path of a dolphin pod by crossing in front of them or cutting through a dolphin pod.

Similarly, Lundquist et al. (2012) found that disruption caused by tour vessel activities disturbed the resting patterns of Dusky dolphins at Kaikoura, a behaviour which has the potential to affect the energetic budget of the dolphins. Moreover, a significant increase in milling behaviour has been observed which may be a result of acoustic masking of communication caused by tour vessels' noise. Acoustic masking of communication can potentially reduce foraging efficiency and expose individuals to a higher predation risk.

8.5 Sperm whales

At Kaikoura, Gordon, Leaper, Hartley and Chappell (1992) found that when tour vessels were present, 'resident' sperm whales had a lower number of ventilations, spent less time on the surface, and were often clearly disturbed by irresponsible vessel handling. Another study conducted by Richter, Dawson and Slooten (2006) reported changes in vocalisation and ventilation patterns of sperm whales when tour vessels were present. The whales displayed a reduction in blow intervals; which may be a stress response. In addition, in the presence of vessels, individuals were found to start echo-locating sooner after starting a dive which may indicate that noise from vessels reduced the effectiveness of the echo-location (Richter et al., 2006).

8.6 New Zealand fur seals

According to Cowling et al. (2015), New Zealand fur seals in the Bay of Plenty, North Island became disturbed when tour vessels approached closer than 20 metres. They rested less and became more alert. Pups, in particular, appeared to be most influenced by vessel distance; they became significantly more alert when a tour vessel was closer and frequently changed their location. As a result, these researchers recommended a 50 metres precautionary minimum approach distance and emphasised the need to develop protocols that mitigate such negative impacts.

In Ohau Stream (25 kilometres north of Kaikoura), Acevedo-Gutierrez, Acevedo and Boren (2011) found that tourists often violated the regulations (MMPR) by harassing (approaching, touching, feeding or throwing objects at) young New Zealand fur seals. Reasons for harassment included a lack of awareness of the regulations, a lack of on-site posted regulations and a lack of enforcement. Acevedo-Gutierrez et al. (2011) suggested posting official-looking volunteers at viewing-sites as an effective and inexpensive tourism-management strategy.

9. Effects of private marine recreationists on marine mammals

Marine mammals' proximity to private recreationists including kayakers, stand-up paddle boarders, surfers and recreational boat users may also have negative effects on marine mammals. Such encounters can cause disturbances to pinnipeds' (Osinga, Nussbaum, Brakefield, & Udo de Haes, 2012) and cetaceans' behaviour (Fandel et al., 2015). For example, studies showed that when kayakers were present, orcas displayed avoidance behaviour, resulting in alterations to time spent feeding. Furthermore, cetaceans that are habituated to recreationists or sought out interactions with them can display aggression towards them (injuries, and in one case human death) or may sustain injuries as a result of boat strikes (Fandel et al., 2015).

In recent years, there are increasing numbers of marine recreationists in New Zealand who encounter and interact with marine mammals (e.g. orcas, seals) both intentionally and incidentally while participating in their chosen recreational activities. In a number of these encounters, marine recreationists were found to violate the regulations (MMPR) (Orams, 2016). However, recent threats from the Department of Conservation (DoC) to prosecute these private recreationists under the MMPA for violating the regulations were met with outrage from many (Orams, 2016), indicating a lack of awareness among marine recreationists and the general public about the regulations and guidelines, and their importance. Thus, both interactions between marine mammals and private recreationists as well as the commercial marine mammal tourism industry are raising increasing concerns about the effects of such activities.

The above-mentioned examples provide evidence of the disturbance of natural behaviour patterns of a variety of marine mammal species caused by tourism and recreational activities in various locations in New Zealand.

Non-compliance with the current regulations (MMPR) has been also documented and there are numerous calls for changes to the current management regime to ensure the sustainability of the industry and conservation of the targeted species.

10. Conclusions

New Zealand is viewed as a world leader in marine mammal conservation and its regime for managing marine mammal tourism has been touted as 'best practice' (Markowitz, Harlin, & Würsig, 1999; Orams, 2004, 2005). Nevertheless, despite the claims of enlightened management, marine mammal tourism in New Zealand presents a challenging tourism-environment management situation (Lundquist, 2014; New et al., 2015; Orams, 2004). New Zealand's *Marine Mammals Protection Act* (MMPA; New Zealand Government, 1978) and associated *Marine Mammals Protection Regulations* (MMPR; New Zealand Government, 1992) which were created to protect and conserve marine mammals (New Zealand Government, 1992; Orams, 2004) and to control the large marine mammal tourism industry in New Zealand have been in effect for a long period of time (over 25 years). A variety of studies conducted in New Zealand (and elsewhere) have shown that in every situation, location and species the current patterns and rates of growth of marine tourism activities have potentially adverse (short-term and long-term) effects on marine mammals (Cowling et al., 2014; Lundquist, 2014; New et al., 2015; Orams, 2004). Moreover, a range of studies have demonstrated that the regulations (MMPR) are frequently violated by marine mammal tour operators in various locations, including, but not limited to, Doubtful Sound (Fiordland), Milford Sound (Fiordland), Bay of Plenty, Akaroa and Ohau Stream, Kaikoura (Acevedo-Gutierrez et al., 2011; Lusseau, 2006; Lusseau et al., 2006; Martinez et al., 2011; Meissner et al., 2015). As a consequence, the sustainability of the industry is under question.

Therefore, there is a need for an investigation and assessment of the effectiveness of the current marine mammal tourism management regime in New Zealand to ensure the long-term sustainability of the industry and conservation of the targeted species.

References

Acevedo-Gutierrez, A., Acevedo, L., Belonovich, O., & Boren, L. (2010). How effective are posted signs to regulate tourism? An example with New Zealand fur seals. *Tourism in Marine Environments, 7*(1), 39–41.

Acevedo-Gutierrez, A., Acevedo, L., & Boren, L. (2011). Effects of the presence of official-looking volunteers on harassment of New Zealand fur seals. *Conservation Biology, 25*(3), 623–627.

Allen, S., Smith, H., Waples, K., & Harcourt, R. (2007). The voluntary code of conduct for dolphin watching in Port Stephens, Australia: Is self regulation an effective management tool? *Journal of Cetacean Research and Management, 9*(2), 159–166.

Allen, S.J. (2014). From exploitation to adoration: The historical and contemporary contexts of human-cetacean interactions. In J. Higham, L. Bejder & R. Williams (Eds.), *Whale-watching: Sustainable tourism and ecological management* (pp. 31–47). Cambridge, England: Cambridge University Press.

Auckland Council. (2014). *Auckland sport & recreation strategic action plan 2014-2024*. Auckland, New Zealand: Author. Retrieved from https://www.aucklandcouncil.govt.nz/plans-projects-policies-reports-bylaws/our-plans-strategies/topic-based-plans-strategies/parks-sports-outdoor-plans/Documents/sport-recreation-strategic-action-plan.pdf

Barr, K., & Slooten, E. (1999). *Effects of tourism on dusky dolphins at Kaikoura*. (Conservation Advisory Science Notes: 229). Wellington, New Zealand: Department of Conservation. Retrieved from http://citeseerx.ist.psu.edu/viewdoc/download?doi=10.1.1.220.9329&rep=rep1&type=pdf.

Beasley, I., Bejder, L., & Marsh, H. (2010). *Dolphin-watching tourism in the Mekong River, Cambodia: A case study of economic interests influencing conservation*. International Whaling Commission Working Paper SC/62/WW4.

Beasley, I., Bejder, L., & Marsh, H. (2014). Cetacean-watching in developing countries: A case study from the Mekong River. In J. Higham, L. Bejder & R. Williams (Eds.), *Whale-watching, sustainable tourism and ecological management* (pp. 307–322). Cambridge, England: Cambridge University Press.

Bejder, L., Dawson, S.M., & Harraway, J.A. (1999). Responses by Hector's dolphins to boats and swimmers in Porpoise Bay, New Zealand. *Marine Mammal Science, 15*(3), 738–750.

Bejder, L., Samuels, A.M.Y., Whitehead, H.A.L., Gales, N., Mann, J., Connor, R., ... & Kruetzen, M. (2006). Decline in relative abundance of bottlenose dolphins exposed to long-term disturbance. *Conservation Biology, 20*(6), 1791–1798.

Boren, L.J., Gemmell, N.J., & Barton, K.J. (2002). Tourist disturbance on New Zealand fur seals (*Arctophalus forsteri*). *Australian Mammalogy, 24*(1), 85–96.

Buckley, R. (2000). Neat trends: Current issues in nature, eco- and adventure tourism. *The International Journal of Tourism Research, 2*(6), 437–444.

Cater, C., & Cater, E. (2007). *Marine ecotourism: Between the devil and the deep blue sea*. Wallingford, England: CABI.

Cater, C., Garrod, B., & Low, T. (Eds.). (2015). *The encyclopedia of sustainable tourism*. Wallingford, England: CABI.

Cisneros-Montemayor, A.M., Sumaila, U.R., Kaschner, K., & Pauly, D. (2010). The global potential for whale watching. *Marine Policy, 34*(6), 1273–1278.

Constantine, R. (1999). *Effects of tourism on marine mammals in New Zealand* (Science for Conservation Report 106). Wellington, New Zealand: Department of Conservation.

Constantine, R. (2001). Increased avoidance of swimmers by wild bottlenose dolphins (*Tursiops truncatus*) due to long-term exposure to swim-with-dolphin tourism. *Marine Mammal Science, 17*(4), 689–702.

Constantine, R., & Bejder, L. (2008). Managing the whale- and dolphin-watching industry: Time for a paradigm shift. In J.E.S. Higham & M. Lück (Eds.), *Marine wildlife and tourism management: Insights from the natural and social sciences* (pp. 321–333). Wallingford, England: CABI.

Constantine, R., Brunton, D.H., & Dennis, T. (2004). Dolphin-watching tour boats change bottlenose dolphin (*Tursiops truncatus*) behaviour. *Biological Conservation, 117*, 299–307.

Cowling, M., Kirkwood, R., Boren, L.J., & Scarpaci, C. (2014). The effects of seal-swim activities on the New Zealand fur seal (*Arctophoca australis forsteri*) in the Bay of Plenty, New Zealand, and recommendations for a sustainable tourism industry. *Marine Policy, 45*, 39–44.

Cowling, M., Kirkwood, R., Boren, L., Sutherland, D., & Scarpaci, C. (2015). The effects of vessel approaches on the New Zealand fur seal (*Arctocephalus forsteri*) in the Bay of Plenty, New Zealand. *Marine Mammal Science, 31*(2), 501–519.

Cross, C. L. (2013). *Queen Charlotte Sound, New Zealand: A habitat for marine mammals* (Internal Report to the Department of Conservation, New Zealand). Retrieved from https://e-ko.nz/uploads/1458080393Cheryl%20Cross%20Dolphin%20Research.pdf

Currey, R.J.C., Dawson, S.M., Slooten, E., Schneider, K., Lusseau, D., Boisseau, O.J., ... Williams, J.A. (2009). Survival rates for a declining population of bottlenose dolphins in Doubtful Sound, New Zealand: An information theoretic approach to assessing the role of human impacts. *Aquatic Conservation: Marine and Freshwater Ecosystems, 19*(6), 658–670.

Dans, S.L., Crespo, E.A., & Coscarella, M.A. (2017). Wildlife tourism: Underwater behavioral responses of South American sea lions to swimmers. *Applied Animal Behaviour Science, 188*, 91–96.

Davis, D., & Gartside, D.F. (2001). Challenges for economic policy in sustainable management of marine natural resources. *Ecological Economics, 36*(2), 223–236.

Department of Conservation. (2020a). Marine mammals. Retrieved from http://www.doc.govt.nz/nature/native-animals/marine-mammals/

Department of Conservation. (2017b). *Sharing our coasts with marine mammals.* Retrieved from http://www.doc.govt.nz/Documents/conservation/native-animals/marine-mammals/marine-mammal-regulations-summary.pdf.

Department of Conservation. (2020b). Sharing our coasts with marine mammals. Retrieved from https://www.doc.govt.nz/nature/native-animals/marine-mammals/sharing-our-coasts-with-marine-mammals/

D'Lima, C. (2015). *Striking a balance between fishing, tourism and dolphin conservation at Chilika Lagoon, India* (Doctoral dissertation). Retrieved from https://researchonline.jcu.edu.au/46578/

Duprey, N.M.T., Weir, J.S., & Würsig, B. (2008). Effectiveness of a voluntary code of conduct in reducing vessel traffic around dolphins. *Ocean & Coastal Management, 51*(8–9), 632–637.

Fandel, A.D., Bearzi, M., & Cook, T.C. (2015). Effects of ocean recreational users on coastal bottlenose dolphins (*Tursiops truncatus*) in the Santa Monica Bay, California. *Bulletin, Southern California Academy of Sciences, 114*(2), 63–75.

Forestell, P.H. (2008). Protecting the ocean by regulating whale watching: The sound of one hand clapping. In J.E.S. Higham & M. Lück (Eds.), *Marine wildlife and tourism management: Insights from the natural and social sciences* (pp. 272–293). Wallingford, England: CABI.

Garrod, B., & Fennell, D.A. (2004). An analysis of whalewatching codes of conduct. *Annals of Tourism Research, 31*(2), 334–352.

Gjerdalen, G., & Williams, P.W. (2000). An evaluation of the utility of a whale watching code of conduct. *Tourism Recreation Research, 25*(2), 27–36.

Gordon, J., Leaper, R., Hartley, F.G., & Chappell, O. (1992). *Effects of whalewatching vessels on the surface and underwater acoustic behaviour of sperm whales off Kaikoura, New Zealand* (Science and Research Series No. 52). Wellington, New Zealand: Department of Conservation.

Hansen, A.S. (2016). *Understanding recreational landscapes: Developing a knowledge base on outdoor recreation monitoring in Swedish coastal and marine areas* (Doctoral dissertation). Retrieved from http://hdl.handle.net/2077/49557

Heenehan, H.L., Van Parijs, S.M., Bejder, L., Tyne, J.A., & Johnston, D.W. (2017). Using acoustics to prioritize management decisions to protect coastal dolphins: A case study using Hawaiian spinner dolphins. *Marine Policy, 75*, 84–90.

Higham, J., Bejder, L., & Williams, R. (2014). Tourism, cetaceans and sustainable development: Moving beyond simple binaries and intuitive assumptions. In J. Higham, L. Bejder & R. Williams (Eds.), *Whale-watching: Sustainable tourism and ecological management* (pp. 1–15). Cambridge, England: Cambridge University Press.

Higham, J.E.S., Bejder, L., Allen, S.J., Corkeron, P.J., & Lusseau, D. (2016). Managing whale-watching as a non-lethal consumptive activity. *Journal of Sustainable Tourism, 24*(1), 73–90.

Higham, J.E.S., & Hendry, W.F. (2008). Marine wildlife viewing: Insights into the significance of the viewing platform. In J.E.S. Higham & M. Lück (Eds.), *Marine wildlife and tourism management: Insights from the natural and social sciences* (pp. 347–360). Wallingford, England: CABI.

Higham, J.E.S., & Lück, M. (2008). Marine wildlife and tourism management: In search of scientific approaches to sustainability. In J.E.S. Higham & M. Lück (Eds.), *Marine wildlife and tourism management: Insights from the natural and social sciences* (pp. 1–16). Wallingford, England: CABI.

Higham, J., & Neves, K. (2015). Whales, tourism and manifold capitalist fixes: New relationships with the driving force of capitalism. In K. Markwell (Ed.), *Animals and tourism: Understanding diverse relationships* (pp. 109–127). Bristol, England: Channel View Publications.

Hoyt, E. (2001). *Whale watching 2001 – worldwide tourism numbers, expenditures, and expanding socioeconomic benefits*. Yarmouth Port, MA: International Fund for Animal Welfare (IFAW) and the United Nations Environmental Program (UNEP).

Hoyt, E. (2009). Whale watching. In W.F. Perrin, B. Würsig & J.C.M. Thewissen (Eds.), *Encyclopedia of marine mammals* (pp. 1223–1227). San Diego, CA: Academic Press.

Hoyt, E., & Parsons, E.C.M. (2014). The whale-watching industry: Historical development. In J. Higham, L. Bejder & R. Williams (Eds.), *Whale-watching, sustainable tourism and ecological management* (pp. 57–70). Cambridge, England: Cambridge University Press.

Hughes, P. (2001). Animals, values and tourism – structural shifts in UK dolphin tourism provision. *Tourism Management, 22*(4), 321–329.

International Fund for Animal Welfare (IFAW) and Tethys European Conservation. (1995). *Report of the workshop on the scientific aspects of managing whale-watching*. Montecastello di Vibio, Italy: International Fund for Animal Welfare, Tethys European Conservation.

Jefferies, C.S. (2016). *Marine mammal conservation and the law of the sea*. New York, NY: Oxford University Press.

Johnson, G., & McInnis, C. (2014). An effective education programme is no fluke. In J.E.S. Higham, L. Bejder & R. Williams (Eds.), *Whale-watching: Sustainable tourism and ecological management* (pp. 128-145). Cambridge, England: Cambridge University Press.

Kessler, M., & Harcourt, R. (2010). Aligning tourist, industry and government expectations: A case study from the swim with whales industry in Tonga. *Marine Policy, 34*(6), 1350-1356.

Kirkwood, R., Boren, L., Shaughnessy, P., Szteren, D., Mawson, P., Hückstädt, L., ... Berris, M. (2003). Pinniped-focused tourism in the Southern hemisphere: A review of the industry. In N. Gales, M. Hindell & R. Kirkwood (Eds.), *Marine mammals: Fisheries, tourism, and management issues* (pp. 245-264). Collingwood, Australia: CSIRO Publishing.

Lemelin, R.H., & Dyck, M. (2008). New frontiers in marine wildlife tourism: An international overview of polar bear tourism management strategies. In J.E.S. Higham & M. Lück (Eds.), *Marine wildlife and tourism management: Insights from the natural and social sciences* (pp. 361-379). Wallingford, England: CABI.

Lück, M. (2003). Education on marine mammal tours as agent for conservation – but do tourists want to be educated? *Ocean & Coastal Management, 46*(9-10), 943-956.

Lück, M. (2008a). Managing marine wildlife experiences: The role of visitor interpretation programmes. In J.E.S. Higham & M. Lück (Eds.), *Marine wildlife and tourism management: Insights from the natural and social sciences* (pp. 334-346). Wallingford, England: CABI.

Lück, M. (Ed.). (2008b). *The encyclopedia of tourism and recreation in marine environments*. Wallingford, England: CABI.

Lück, M. (2009). *Environmentalism and tourists' experiences on swim-with dolphins tours: A case study of New Zealand*. Saarbrücken, Germany: VDM-Verlag.

Lück, M. (2015). Education on marine mammal tours – but what do tourists want to learn? *Ocean & Coastal Management, 103*, 25-33. doi:http://dx.doi.org/10.1016/j.ocecoaman.2014.11.002

Lück, M., & Higham, J.E.S. (2008). Marine wildlife and tourism management: Scientific approaches to sustainable management. In J.E.S.

Higham & M. Lück (Eds.), *Marine wildlife and tourism management: Insights from the natural and social sciences* (pp. 380–388). Wallingford, England: CABI.

Lundquist, D. (2014). Management of dusky dolphin tourism in Kaikoura (New Zealand). In J.E.S. Higham, L. Bejder & R. Williams (Eds.), *Whale-watching: Sustainable tourism and ecological management* (pp. 337–351). Cambridge, England: Cambridge University Press.

Lundquist, D., Gemmell, N., & Würsig, B. (2012). Behavioural responses of dusky dolphin groups (*Lagenorhynchus obscurus*) to tour vessels off Kaikoura, New Zealand. *PLoS ONE, 7*(7), e41969.

Lusseau, D. (2006). The short-term behavioral reactions of bottlenose dolphins to interactions with boats in Doubtful Sound, New Zealand. *Marine Mammal Science, 22*(4), 802–818.

Lusseau, D. (2008). Understanding the impacts of noise on marine mammals. In J.E.S. Higham & M. Lück (Eds.), *Marine wildlife and tourism management: Insights from the natural and social sciences* (pp. 206–218). Wallingford, England: CABI.

Lusseau, D., & Higham, J.E.S. (2004). Managing the impacts of dolphin-based tourism through the definition of critical habitats: The case of bottlenose dolphins (*Tursiops spp.*) in Doubtful Sound, New Zealand. *Tourism Management, 25*(6), 657–667.

Lusseau, D., Slooten, L., & Currey, R.J.C. (2006). Unsustainable dolphin-watching tourism in Fiordland, New Zealand. *Tourism in Marine Environments, 3*(2), 173–178.

Markowitz, T., Harlin, A., & Würsig, B. (1999). *New Zealand dusky dolphins*. Laredo, Texas and Kaikoura, New Zealand: Earthwatch & Department of Wildlife and Fisheries Science, Texas A&M University.

Martinez, E., Orams, M.B., & Stockin, K.A. (2011). Swimming with an endemic and endangered species: Effects of tourism on Hector's dolphins in Akaroa Harbour, New Zealand. *Tourism Review International, 14*, 99–115.

Meissner, A.M., Christiansen, F., Martinez, E., Pawley, M.D.M., Orams, M.B., & Stockin, K.A. (2015). Behavioural effects of tourism on oceanic common dolphins, *Delphinus sp.*, in New Zealand: The effects of Markov analysis variations and current tour operator compliance with regulations. *PLoS ONE, 10*(1), e0116962.

Mustika, P.L.K. (2011). *Towards sustainable dolphin watching tourism in Lovina, Bali, Indonesia* (Doctoral dissertation). Retrieved from https://researchonline.jcu.edu.au/29750/

Mustika, P.L.K., Birtles, A., Everingham, Y., & Marsh, H. (2013). The human dimensions of wildlife tourism in a developing country: Watching spinner dolphins at Lovina, Bali, Indonesia. *Journal of Sustainable Tourism, 21*(2), 229–251.

Mustika, P.L.K., Birtles, A., Everingham, Y., & Marsh, H. (2015). Evaluating the potential disturbance from dolphin watching in Lovina, north Bali, Indonesia. *Marine Mammal Science, 31*(2), 808–817.

Mustika, P.L.K., Birtles, A., Welters, R., & Marsh, H. (2012). The economic influence of community-based dolphin watching on a local economy in a developing country: Implications for conservation. *Ecological Economics, 79*, 11–20.

Mustika, P.L.K., Welters, R., Ryan, G.E., D'Lima, C., Sorongon-Yap, P., Jutapruet, S., & Peter, C. (2017). A rapid assessment of wildlife tourism risk posed to cetaceans in Asia. *Journal of Sustainable Tourism, 25*(8), 1138–1158.

Neumann, D.R., & Orams, M.B. (2006). Impacts of ecotourism on short-beaked common dolphins (*Delphinus delphis*) in Mercury Bay, New Zealand. *Aquatic Mammals, 32*(1), 1–9.

New protection for bottlenose dolphin. (2019, August 29). Retrieved from https://www.doc.govt.nz/news/media-releases/2019/new-protection-for-bottlenose-dolphin/

New, L.F., Hall, A.J., Harcourt, R., Kaufman, G., Parsons, E.C.M., Pearson, H.C., ... Schick, R.S. (2015). The modelling and assessment of whale-watching impacts. *Ocean & Coastal Management, 115*, 10–16.

Newsome, D., & Rodger, K. (2008). Impacts of tourism on pinnipeds and implications for tourism management. In J.E.S. Higham & M. Lück (Eds.), *Marine wildlife and tourism management: Insights from the natural and social sciences* (pp. 182–205). Wallingford, England: CABI.

New Zealand Government. (1978). *New Zealand marine mammals protection act.* Wellington, New Zealand: Government Printer. Retrieved from http://www.legislation.govt.nz/act/public/1978/0080/latest/DLM25111.html.

New Zealand Government. (1991). *Resource management act.* Wellington, New Zealand: Government Printer. Retrieved from http://www.legislation.govt.nz/act/public/1991/0069/211.0/DLM230265.html

New Zealand Government. (1992). *New Zealand marine mammals protection regulations.* Wellington, New Zealand: Government Printer. Retrieved from http://www.legislation.govt.nz/regulation/public/1992/0322/latest/DLM168286.html

O'Connor, S., Campbell, R., Cortez, H., & Knowles, T. (2009). *Whale watching worldwide: Tourism numbers, expenditures and expanding economic benefits.* Yarmouth, MA: International Fund for Animal Welfare.

Orams, M. (1994). Creating effective interpretation for managing interaction between tourists and wildlife. *Australian Journal of Environmental Education, 10*, 21–34.

Orams, M.B. (1996). A conceptual model of tourist–wildlife interaction: The case for education as a management strategy. *The Australian Geographer, 27*(1), 39–51.

Orams, M.B. (1997). The effectiveness of environmental education: Can we turn tourists into 'greenies'? *Progress in Tourism and Hospitality Research, 3*(4), 295–306.

Orams, M. (1999). *Marine tourism: Development, impacts and management.* London, England: Routledge.

Orams, M.B. (2000). Tourists getting close to whales, is it what whale-watching is all about? *Tourism Management, 21*(6), 561–569.

Orams, M. (2004). Why dolphins may get ulcers: Considering the impacts of cetacean-based tourism in New Zealand. *Tourism in Marine Environments, 1*(1), 17–28.

Orams, M. (2005). Dolphins, whales and ecotourism in New Zealand: What are the impacts and how should the industry be managed? In C.M. Hall & S. Boyd (Eds.), *Nature-based tourism in peripheral areas: Development or disaster* (pp. 231–245)? Clevedon, England: Channel View Publications.

Orams, M. (2016, September, 15). *Why we shouldn't swim with marine mammals.* Retrieved from http://www.nzherald.co.nz/opinion/news/article.cfm?c_id=466&objectid=11709601

Orams, M., Forestell, P., & Spring, J. (2014). What's in it for the whales? In J. Higham, L. Bejder & R. Williams (Eds.), *Whale-watching: Sustainable tourism and ecological management* (pp. 146–162). Cambridge, England: Cambridge University Press.

Osinga, N., Nussbaum, S.B., Brakefield, P.M., & Udo de Haes, H.A. (2012). Response of common seals (*Phoca vitulina*) to human disturbances in the Dollard estuary of the Wadden Sea. *Mammalian Biology, 77*(4), 281–287.

Parsons, E.C.M. (2012). The negative impacts of whale-watching. *Journal of Marine Biology, 2012*, 1–9.

Peters, C., & Stockin, K. (2016). *Responses of bottlenose dolphins (Tursiops truncatus) to vessel activity in Northland, New Zealand* (Progress report). Northland, New Zealand: Department of Conservation.

Piper, D. (2019, October 02). *DOC says dolphins come first as swimming ban hits Bay of Islands tour operators.* Retrieved from https://www.stuff.co.nz/business/116262180/doc-says-dolphins-come-first-as-swimming-ban-hits-bay-of-islands-tour-operators

Richter, C., Dawson, S., & Slooten, E. (2006). Impacts of commercial whale watching on male sperm whales at Kaikoura, New Zealand. *Marine Mammal Science, 22*(1), 46–63. doi: 10.1111/j.1748-7692.2006.00005.x

Shawky, A.M., & Afifi, A. (2008). Behaviour of spinner dolphin at Sha'ab Samadai, Marsa Alam, Red Sea, Egypt. *Egyptian Journal of Biology, 10*(1), 36–41.

Shultis, J. (2012). The impact of technology on the wilderness experience: A review of common themes and approaches in three bodies of literature. In D.N. Cole (Ed.), *Wilderness visitor experiences: Progress in research and management; 2011 April 4-7; Missoula, MT. Proc. RMRS-P-66* (pp. 110–118). Fort Collins, CO: U.S. Department of Agriculture, Forest Service, Rocky Mountain Research Station.

Shultis, J.D. (2015). "Completely empowering": A qualitative study of the impact of technology on the wilderness experience in New Zealand. In A. Watson, S. Carver, Z. Křenová & B. McBride (Eds.), *Science and stewardship to protect and sustain wilderness values: Tenth world wilderness congress symposium; 2013, 4-10 October, Salamanca, Spain. Proceedings RMRS-P-74.* (pp. 195–201). Fort Collins, CO: U.S. Department of Agriculture, Forest Service, Rocky Mountain Research Station.

Sorice, M.G., Shafer, C.S., & Ditton, R.B. (2006). Managing endangered species within the use–preservation paradox: The Florida manatee (*Trichechus manatus latirostris*) as a tourism attraction. *Environmental Management, 37*(1), 69–83.

Sun, X., & Lück, M. (2015). The internet presence of whale and dolphin watch operators in New Zealand in terms of their ecotourism attributes: A content analysis. In M. Lück, J. Velvin & B. Eisenstein (Eds.), *The social side of tourism: The interface between tourism, society, and the environment* (pp. 141–156). Frankfurt, Germany: Peter Lang Verlag.

Timmel, G., Courbis, S., Sargeant-Green, H., & Markowitz, H. (2008). Effects of human traffic on the movement patterns of Hawaiian spinner dolphins (*Stenella longirostris*) in Kealakekua Bay, Hawaii. *Aquatic Mammals, 34*(4), 402–411.

United Nations. (2018). *The sustainable development agenda.* Retrieved from https://www.un.org/sustainabledevelopment/development-agenda/

Young, K. (1998). *Seal watching in the UK and Republic of Ireland.* London, England: International Fund for Animal Welfare, UK.

Zeppel, H., & Muloin, S. (2008). Marine wildlife tours: Benefits for participants. In J.E.S. Higham & M. Lück (Eds.), *Marine wildlife and tourism management: Insights from the natural and social sciences* (pp. 19–48). Wallingford, England: CABI.

Zeppel, H., & Muloin, S. (2014). Green messengers or nature's spectacle: Understanding visitor experiences of wild cetacean tours. In J. Higham, L. Bejder & R. Williams (Eds.), *Whale-watching: Sustainable tourism and ecological management* (pp. 110–126). Cambridge, England: Cambridge University Press.

Eva Holmberg

It is mostly about money- discussions related to the political decision-making resulting in giant pandas moving into Ähtäri Zoo

1. Introduction

Animals and their role in tourism have been discussed rather extensively, from cloning animals for tourism (Wright, 2018), human-animal relationships (Markwell, 2015), animal welfare (Hughes, 2001), zoos (Frost, 2010) all the way to how animals in general can be used in tourism (Äijälä, Carcia-Rosell & Haanpää, 2016). The importance of animal welfare is an issue that has been raised above the others, especially by Fennell (2015; 2012). Maybe somewhat surprisingly, very few studies have focused on giant pandas (*Ailuropoda melanoleuca*) being borrowed from China for becoming tourism attractions in countries chosen by the Chinese government.

The giant panda is well known for its cuteness as well as being a symbol for the World Wildlife Fund. The giant panda was close to extinction some decades ago, but in 2016 the status of the panda was downgraded by the International Union for Conservation of Nature (IUCN), from endangered to vulnerable, i.e. the work to save the species has been successful. Pandas were used for instance in social media by the Chinese government to improve its image, but some selected countries have also been honoured by receiving pandas for a certain amount of time (Anderlini, 2017). For centuries Chinese leaders have used pandas as a tool for enhancing its international relationships, a strategy which has been called panda diplomacy (Szczudlik, 2017). It is though not only the Chinese government using pandas for policy making, when a country accepts the pandas the bears become a political issue in the receiving country.

In 2016, the Finnish Ministry of Agriculture and Forestry made the decision to accept a panda couple offered by the Chinese Government as a recognition to Finland's 100 years of independence in 2017 (Harald, 2017). Later that year it was decided that the pandas will be placed in Ähtäri, a municipality with 6000 inhabitants located 350 kilometres from the capital area. Ever since then, the pandas have been presented as a unique tourism attraction to convince local people of the value of the investment. Local people are worried about the costs of the new panda house required by the Chinese authorities as well as about the estimated annual fee of 1 million euros for rent of the pandas. The decision to accept the pandas and the decision to locate them in Ähtäri has raised a lot of discussion by media representatives, tourism experts and the Finnish public. This paper will take a closer look to identify how political decision-making at different geographical levels resulted in a panda couple moving into a Finnish zoo with a rather peripheral location. Considering the important role of governments in tourism development (see e.g.Pastras & Bramwell, 2013; Hall & Jenkins, 2004), the aim of this study is to explore how the controversial political decisions related to accepting the pandas to Finland and Ähtäri Zoo were discussed in Finnish media. The study was conducted by analysing newspaper and Internet articles published from the beginning of 2014 until end of July 2018.

2. Current political agendas in Finland

Policymaking takes place in different contexts from nations to organisations (Hall, 2004). Policymaking is "a definite course or method of action selected from among alternatives and in light of given conditions to guide and determining present and future actions" (Edgell & Swanson, 2013, p. 30). Governments, involving formal and informal institutions and practices at hierarchical levels from national, regional to local, are public sector actors involved in the activity (Pastras & Bramwell, 2013). Hall (2010, p. 8) argues that public policy is "whatever officials within governments choose to do or not to do about issues that require government intervention". Policy is not limited to a written document, it also covers issues such as the purpose of governmental action and goals to be achieved (Hall, 2010). Thus, policy is about making plans and taking actions enhancing the wellbeing of the society (Sojern, 2005; Hall & Jenkins, 2004). It is about making decisions influencing both internal and international matters. It is important that governments are successful in creating trust among the citizens, citizens need to be able to count on that the political system produces preferred outcomes. (Pagilara, Asia, Russo, Della Corte & Nunko, 2020)

The Finnish Government 2015–2019 stated in its initial political program (from 2015) that the aim of the years to come is to strengthen Finland's position globally, secure the independence, stay neutral (no NATO membership), and guarantee the wellbeing and safety of Finnish citizens. In a world built on interdependencies, Finland will enhance international stabilization, peace, democracy, human rights, good political leadership, sustainable development and equality. (Program of Government, 2015). The political program from 2015 identifies bio economy, forestry, farming and circular economy as key industries in Finland in the years to come. The service sector including tourism and hospitality was not mentioned in the original program. After two years, the government launched a new program that omprises a project called Tourism 4.0 aiming at giving funding for digitalization, intensified marketing and measures taken for decreasing the challenge of seasonality (Kasvun kärki matkailu 4.0, 2017).

Overall, governments have often been criticised for implementing tourism policies that are short-term and lack a clear direction. (Pagilara et al., 2020). According to Ritchie and Crouch (2003), tourism policy is simply a way to define the rules of the game but these authors also provide a more comprehensive definition:

> *"Tourism policy is a set of regulations, rules, guidelines, directives and development/ promotion objectives and strategies that provide a framework within which the collective and individual decisions directly affecting tourism development and the daily activities are taken (Ritchie & Crouch 2003, p.148)*

The current tourism policy of Finland is summarized in a tourism strategy that in Finland is referred to as the Tourism Road Map 2015–2025. The aim of the roadmap is to make Finland the most attractive destination in Northern Europe by 2025. The road map presents a joint vision and mission as well as identifies some key strategies to help them to achieve the vision. The key focus is on strengthening business and destination networks, development of interesting supply including products and services like Outdoors Finland, Culture Finland, water tourism, wellness and health services, physical exercises and sports, food production, national parks and cultural heritage (Roadmap, 2015), marketing actions and creation of a competitive operational environment. Finland's strengths in the road map are listed as: unpolluted nature, circumstances enhancing wellbeing, original culture and well-functioning infrastructure. Authenticity (Suomen matkailun, 2014) and responsibility (Teemoista sisältöä, 2020) are important starting points in all tourism development work.

3. Panda diplomacy

Panda (*Ailuropoda melanoleuca*) is a species of bear and it is an animal linked to a lot of symbolism. The World Wildlife Fund for Nature (WWF) has used the panda as its logo since 1961 and, in China pandas are considered as a representative for the unique culture of the country. (Anderlini, 2017) At the end of the 1980s, the panda was near extinction due to the human population growth, farming and demolishment of the bamboo forests in China (Högmander & Vanhalakka, 2018). The panda population is still small, although it has slightly increased due to panda cubs being artificially bred at research centres in China (Hogenboom, 2013). Today, only some 1,800 to 2,000 animals live in the wild and 300 in zoos (Szczudlik, 2017; Siltamäki, 2018a).

According to Harvard professor Joseph Nye, a country can improve its international image in three ways; through its culture, through its political institutions or by its foreign policy (Andrelini, 2017). For more than 50 years China has used pandas to improve its image and enhance its relationships with other countries. This began during the years under Mao Zedong when it was realised that the giant panda could become a national treasure and support cooperation with other countries. In the beginning, pandas were given as a gift, but later on the pandas were sent to foreign countries as loans and the receiving country is now forced to pay rent and usually also expecting to supply China with new technology and knowledge in exchange (Hogenbook, 2013).

The amount the zoos receiving the pandas have to pay to China has not be disclosed to the public in Finland, but it is estimated to be around 1 million euros a year, i.e. the same amount as for instance Berlin's Tierpark pays. The Chinese government argues that 70 percent of the money China receives from the rent of the pandas go to the conservation of the panda stock in China, but there is no transparency of how the money is used (Anderlini, 2017).

In 2017, up to 70 pandas had been loaned to twenty countries, and negotiations for more pandas to move abroad is continuing all the time. Apparently,the purpose of the business has also changed:In the beginning it was more about getting help with the conservation of the panda stock, but now the main purpose appears to be political. (Anderlini, 2017) The current Chinese president Xi uses pandas to enhance a softer image of China and he is personally signing every loan of pandas (Szczudlik, 2017; Anderlini, 2017). The signing of the agreements has often taken place at the same time as major trade deals have been signed between China and the receiving country. For instance, Australia and France agreed upon selling nuclear technology to China at the same time as the move of the pandas to zoos in these countries was discussed. Thus, it can be stated that, from a Chinese perspective, the most important trading partners have received pandas as a sign of long-term relationships categorised by trust (Anderlini, 2017; Karjalainen, 2017; Szczudlik, 2017).

4. Ähtäri Zoo

Human beings have been interested in interacting and seeing animals since ancient times, and today animals are an important travel motivator today. Thus, animals have been commodified into tourism products all over the

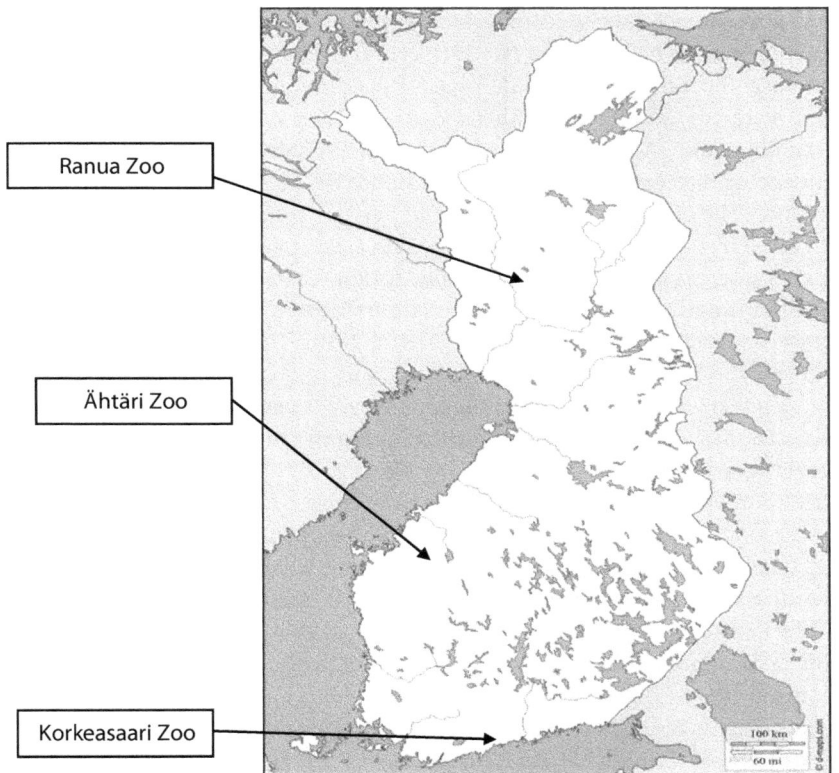

Figure 1. Location of Finland's most popular zoos (modified from Finland, 2020)

world. This commodification can take place in the wild or in zoos. (Markwell, 2015) The commodification of animals for tourists in Finland has mostly taken place in zoos, even if products such as bear watching and seal safaris in the natural environment exist as well. Finland has three bigger zoos; Korkeasaari in Helsinki, Ranua Zoo in Lapland and Ähtäri in South Ostrobothnia (Figure 1). Korkeasaari attracts more than 500,000 visitors annually, Ähtäri 170,000 and Ranua almost 160,000.

As a municipality Ähtäri has a long history as an industrial village, with a water powered saw mill and an ironworks, both established in the 19th century. By the 1960s, the glory days of the municipality were over, and its inhabitants left the rural area for bigger cities. In order to avoid this exodus, the head of the local government suggested that Ähtäri could become a tourism destination just as it used to be in the the late 19th century, when for instance the Finnish composer Sibelius travelled to the area for inspiration. A zoo was chosen as the main attractor (Siltamäki, 2018b).

Subsequently, Ähtäri Zoo was founded in 1973 during the global oil crisis influencing the economic situation in Finland (Helin, 2018). Its first animal was a moose. The aim was that the zoo would attract tourists and their spending, enhanced by the multiplier effect, that would create jobs and income for a municipality in crisis (Helin, 2018; Siltamäki, 2018b). The zoo was concentrating on Nordic animals. Even though the animals are kept in captivitym the cages are extensive and as natural as possible (Helin, 2018). The zoo has more than 60 different species including wild boar, lynx, wolf and brown bear (Ähtäri, 2018). The zoo is owned by the Ähtäri municipality (99 percent), but it has its own company, and thereby its own budget resulting in that the zoo at all the times should be profitable (Siltamäki, 2018a). Nearby, the zoo is supported by some basic tourism infrastructure and services such as a hotel, a camping site and some rental cottages, a golf course and an adventure park built in a natural setting (Ähtäri Zoo, 2018). When Ähtäri became the home for the panda couple, it was after several years of lobbying at different political levels.

The final agreement on the pandas was signed in April 2017 when the president of Finland Sauli Niinistö met his Chinese counterpart Xi Jinping (Karjalainen, 2017). The process to get the pandas to Finland required efforts from different Finnish authorities such as ministries, the South Ostrobothnia regional council and local decision-makers in Ähtäri. The pandas Lumi (Jin Bao Bao) and Pyry (Hua Bao) (Figure 2) arrived from China in January 2018, and in February the panda house was opened to the public (Huhtanen, 2018; Högmander & Vanhalakka, 2018).

The fact that Ähtäri municipality owns 99 percent of the zoo has raised a lot of concerns among local people when in 2016 it become clear that Ähtäri Zoo would be the zoo receiving the panda bears from China. The concern was justified, year 2017 resulted in deficit of about 1 million euros for the company running the zoo (Forsman, 2018). In order to reach a break even result, the pandas must attract 100,000 more visitors than in the previous years, when the number of visitors has been around 160,000 to 170,000 (Högmander & Vanhalakka, 2018).

The Panda house at Ähtäri Zoo can accommodate 5,400 daily visitors and the ticket price is 35 euros per person for access to the panda house only, and 39 euros for the entire zoo. The pandas have their own rest areas, climbing facilities and water elements. Visitors can follow the life of the pandas through glass

Figure 2. Pyry in the panda house (picture of author)

walls. The main food of the pandas is bamboo; they need 12 to 30 kg of bamboo a day (Öberg, 2018). The bamboo is brought to Finland weekly by a truck from the Netherlands at an estimated cost of 100,000 euros annually (Karjalainen, 2018). Pandas are picky eaters, and the truck to Ähtäri brings in 8 to12 different bamboo varieties. A selection of different bamboo is supplied for the pandas to eat week to week. (Högmander & Vanhalakka, 2018).

The pandas are in Ähtäri on a loan from China for 15 years (Ähtärin pandatalo avasi, 2018; Karjalainen, 2018). If the pandas were to get a panda cub one day it would be owned by the State of China, and the aim would be that, one day, the cub could return to China to live freely. (Högmander & Vanhalakka, 2018; Siltamäki, 2018a). The next part of this chapter will discuss the topic panda diplomacy and how pandas became a key political tool for China.

5. Ähtäri Zoo

Human beings have been interested in interacting and seeing animals since ancient times, and today animals are an important travel motivator today. Thus, animals have been commodified into tourism products all over the world. This commodification can take place in the wild or in zoos (Markwell, 2015). The commodification of animals for tourists in Finland has mostly taken place in zoos, even if products such as bear watching and seal safaris in the natural environment exist as well. Finland has three bigger zoos; Korkeasaari in Helsinki, Ranua Zoo in Lapland and Ähtäri in South Ostrobothnia (Figure 1). Korkeasaari attracts more than 500,000 visitors annually, Ähtäri 170,000 and Ranua almost 160,000.

As a municipality Ähtäri has a long history as an industrial village, with a water powered saw mill and an ironworks, both established in the 19th century. By the 1960s, the glory days of the municipality were over, and its inhabitants left the rural area for bigger cities. In order to avoid this exodus, the head of the local government suggested that Ähtäri could become a tourism destination just as it used to be in the the late 19th century, when for instance the Finnish composer Sibelius travelled to the area for inspiration. A zoo was chosen as the main attractor (Siltamäki, 2018b).

Subsequently, Ähtäri Zoo was founded in 1973 during the global oil crisis influencing the economic situation in Finland (Helin, 2018). Its first animal was a moose. The aim was that the zoo would attract tourists and their spending, enhanced by the multiplier effect, that would create jobs and income for a municipality in crisis (Helin, 2018; Siltamäki, 2018b). The zoo was concentrating on Nordic animals. Even though the animals are kept in captivitym the cages are extensive and as natural as possible (Helin, 2018). The zoo has more than 60 different species including wild boar, lynx, wolf and brown bear (Ähtäri, 2018). The zoo is owned by the Ähtäri municipality (99 percent), but it has its own company, and thereby its own budget resulting in that the zoo at all the times should be profitable (Siltamäki, 2018a). Nearby, the zoo is supported by some basic tourism infrastructure and services such as a hotel, a camping site and some rental cottages, a golf course and an adventure park built in a natural setting (Ähtäri Zoo, 2018). When Ähtäri became the home for the panda couple, it was after several years of lobbying at different political levels.

The final agreement on the pandas was signed in April 2017 when the president of Finland Sauli Niinistö met his Chinese counterpart Xi Jinping (Karjalainen, 2017). The process to get the pandas to Finland required efforts from different Finnish authorities such as ministries, the South Ostrobothnia regional council and local decision-makers in Ähtäri. The pandas Lumi (Jin Bao Bao) and Pyry (Hua Bao) (Figure 2) arrived from China in January 2018,and in February the panda house was opened to the public (Huhtanen, 2018; Högmander & Vanhalakka, 2018).

The fact that Ähtäri municipality owns 99 percent of the zoo has raised a lot of concerns among local people when in 2016 it become clear that Ähtäri Zoo would be the zoo receiving the panda bears from China. The concern was justified, year 2017 resulted in deficit of about 1 million euros for the company running the zoo (Forsman, 2018). In order to reach a break even result, the pandas must attract 100,000 more visitors than in the previous years, when the number of visitors has been around 160,000 to 170,000 (Högmander & Vanhalakka, 2018).

The Panda house at Ähtäri Zoo can accommodate 5,400 daily visitors and the ticket price is 35 euros per person for access to the panda house only, and 39 euros for the entire zoo. The pandas have their own rest areas, climbing facilities and water elements. Visitors can follow the life of the pandas through glass walls. The main food of the pandas is bamboo; they need 12 to 30 kg of bamboo a day (Öberg, 2018). The bamboo is brought to Finland weekly by a truck from the Netherlands at an estimated cost of 100,000 euros annually (Karjalainen, 2018). Pandas are picky eaters, and the truck to Ähtäri brings in 8 to12 different bamboo varieties. A selection of different bamboo is supplied for the pandas to eat week to week. (Högmander & Vanhalakka, 2018).

The pandas are in Ähtäri on a loan from China for 15 years (Ähtärin pandatalo avasi, 2018; Karjalainen, 2018). If the pandas were to get a panda cub one day it would be owned by the State of China, and the aim would be that, one day, the cub could return to China to live freely. (Högmander & Vanhalakka, 2018; Siltamäki, 2018a). The next part of this chapter will discuss the topic panda diplomacy and how pandas became a key political tool for China.

6. Research methods

According to Veal (2011) analysing texts is an increasingly recognised approach in tourism research. Text data can for instance be reports, minutes of meetings, diaries or newspaper articles (Saunders, Lewis & Thornhill, 2009). The data analysed for this study comprise 26 articles published on the Internet by Finnish media channels during the years 2014, 2016, 2017, and 2018 (Table 1). Most articles are from main Finnish newspapers such as Helsingin Sanomat, Iltalehti, Aamulehti and the Swedish Hufvudstadsbladet. Other sources are media organisations such as the national broadcasting company YLE and one of the commercial broadcasting companies, MTV. The articles were mainly found through Google with key words such as Ähtäri + panda, Finland + panda. Hundreds of articles about the pandas coming to Finland were found, most of them offered only an overview of the life giant pandas in general, what they eat, and how they have adapted to their new home. The 26 articles chosen for this analysis described at least to some extent also the process of how the pandas ended up in Ähtäri and decision making behind the process.

Table 1. The analysed articles

Article	Title of article	Publisher
1	Ähtärin Pandatalolla lupaava alku- Pandoja on tultu katsomaan sankoin joukoin parjatusta lippuhinnasta huolimatta	Iltalehti
2	Ähtärin eläinpuisto vaikeuksissa	MTV
3	Kinas hundraårspresent till Finland: ett pandapar.	Hufvudstadsbladet.
4	Finland fick pandor- mycket oklart kring deras kommande hem.	Hufvudstadsbladet.
5	Ähtäri tarvitse 100 000 panda turistia.	Helsingin Sanomat
6	Ähtärin pandat: Lumi ja Pyry ovat nyt Suomessa-Lue tästä kaikki, mitä sinun pitää pandoista tietää.	Aamulehti
7	Lumi ja Pyry pääsivät telmimään lumessa- pandatalo avautui vihdoin yleisölle (2018)	Iltalehti
8	Pandat pehmentävät liiaksi kuvaa Kiinasta.	Iltasanomat
9	Ähtärin pandojen vuosihinta lähes miljoona euro.	Iltasanomat
10	Ähtärin eläinpuiston pandatalo avasi ovensa: "Pandatalon rakentaminen oli ponnistus ja mittava investointi".	Maaseudun Tulevaisuus
11	Neljä kuntaa täyttää kriisikuntakriteerit (2017)	Kuntalehti
12	Pandat ovat saapuneet Suomeen.	Iltalehti
13	Analyysi: Suomi pääsi Kiinan syleilyyn-Pandamaat eivät mesoa ihmisoikeuksista.	YLE
14	Pelastavatko Pandat Ähtärin?	Aamulehti
15	Kina tar bättre hand om sina pandor än oliktänkare.	Österbottens-tidningen
16	Näkökulma: Pandat ovat söpömpiä kuin ihmisoikeudet.	Iltalehti
17	Ähtärin pandojen vuokrahinta yli 12 miljoona euroa.	Iltalehti
18	Pandoilla pitäisi parantaa kiinalaisten eläinoikeuksia.	Aamulehti
19	Ministeriö haluaa pandat Ähtäriin- lappilaiset pettyivät.	YLE
20	Missä menee pandojen näkemisen kipuraja	Iltalehti
21	Rankkaa kritiikkiä pandojen sijoittamisesta Ähtäriin HS:ssä: " Suuri virhe	Iltalehti
22	Pörröiset turistihoukuttimet voivat kaatua veromaksajien harteille.	Iltalehti
23	Suurlähettiläät	Helsinging Sanomat-Kuukausiliite
24	Ähtäri odottaa pandaturisteja	Kauppalehti
25	Pandemi i vårt land	HBL

Table 1. Continued

Article	Title of article	Publisher
26	Pandat vuokrataan maille, joiden katsotaan suhtautuvan myötämielisesti Kiinaan- Eläimet saapuvat ensi viikolla Suomeen	Aaamulehti

The analysis of the articles was done manually, the first phase being reading the articles several times (see e.g. Veal, 2011). During this process, the key themes/discussions emerged.

7. Findings

This chapter presents the three themes that appeared as crucial when the phenomenon of political decision making during the process of receiving the pandas was analysed. The first is the implications of a Finnish international political agenda, the second is related to how the location of the pandas is linked to national politics and Finnish tourism policy, and the last to how local political decision makers were willing to take the risk of investing in the pandas with money raised by the tax payers.

International politics: Finnish-Chinese relationships

During the process of getting the pandas to Finland, there were significant political issues nvolved. The process began as early as in 2014, and the final agreement was signed in 2017 when the Finnish president Niinistö and Chinese president Xi met in Helsinki. The Finnish president stated during meeting visit that Finland and China have an excellent relationship and that the pandas will further strengthen this (Siltamäki, 2018a). However, there were questions what happened to the governmental program stressing that Finland internationally strive for enhancing democracy, human rights and equality.

Several articles in the main Finnish media channels discuss the fact that China gives pandas to countries that do not criticise it about human rights and have a positive attitude towards China (e.g. Parkkonen, 2018; Buchert, 2018; Mäkeläinen, 2018; Ådahl, 2018). For instance, Mäkeläinen (2017) highlights in an article on the main national broadcasting company YLE 's webpage that it is not only the Finnish climate that is well suitable for the pandas, it is the political climate as well. Mäkeläinen discusses also the fact that the panda diplomacy is a unique strategy for China to improve its image and enhance its power to influence global issues. The Minister of Agriculture and Forestry, Jari Leppä praised the good relationships during a big event organised when the pandas landed in Finland. The minister stated that Finland could not receive a better gift from China (Parkkonen, 2018; Buchert, 2018). What the minister seems to have forgotten is that the pandas are not a gift, but a million-dollar business as the receiving country pays rent for the pandas. According to the minister, the human right issue is a completely different issue that was raised in discussions with other Chinese authorities. The fact is, that China is an important trade partner for Finland and that there are several ongoing industrial projects with China. (Parkkonen, 2018)

None of the articles analysed take the stance that Finland should not have accepted the "gift", even though one anonymous editorial of evening newspaper Iltasanomat published in 2017 highlights the fact that the pandas soften the image of China too much. It seems to be widely accepted that a good relationship with China is crucial, as the country is the fifth most important country for Finnish exports. As Parkkonen (2017) highlights, trade is more important than the tortured journalists and opposition politicians in Chinese jails. Thus, the fact that China has problems with human rights and democracy is acknowledged. However, in general, media representatives do not demand Finnish politicians to act to influence or raise criticism against the situation, which results in the decision to accept the pandas not being criticised.

One article discusses the fact that the pandas coming to Finland could be used by politicians for improving the animal welfare of other animals in Chinese zoos and other places where animals are used, such as in shopping malls. In China animals have only value as objects, to make money with, attracting customers or to give as gifts to friends (Pitkänen, 2017). This is an important discussion since animals, and especially zoos, as tourism attractions, are an important part of the tourism system of many destinations. However, many countries still lack regulations related to how animals can be used in tourism (Äijälä et al., 2016), and the situation in China is poorer than in many other countries. As Pitkänen (2017) says:

> "If the human right situation in China doesn't correspond to the level of the western world, the situation when it comes to animal rights is at the level of the Roman time circus and gladiator shows".

Only a couple of articles take the angle that the pandas in Ähtäri are not so much about politics, but deepening Finnish-Chinese relationships to enhance trade. These articles refer to interviews with the Finnish panda expert Jukka Salo. Salo was involved in the panda project from its beginning, and he is stressing that the panda project of Ähtäri was foremost a project aiming at saving the panda from extinction and not political or related to trade (e.g. Högmander & Vanhalakka, 2018; Koivula, 2018).

National politics

The decision to allocate the pandas to Ähtäri Zoo was made by the Finnish Ministry of Agriculture and Forestry in 2014 (Ådal, 2018; Puoskari, 2014). Of the other Finnish zoos, only Ranua Zoo was interested in giving the pandas a new home. Korkeasaari Zoo in Helsinki decided that the project would become too expensive. (Westerlund, 2018) According to the analysed articles, the reason for choosing Ähtäri Zoo was that Ähtäri was better prepared for the project. The person involved in the negations representing the Ranua zoo stated that they stressed their ability to offer the pandas a good home and good care, whereas Ähtäri Zoo highlighted the business perspective. Even the idea that the marketing the pandas could be combined with marketing Lapland as the home of the Santa Claus to both domestic and international tourists did not result in a decision in favour of Ranua Zoo. (Puoskari, 2014). The climate in Ähtäri, with its four season is similar to the climate in the Sichuan mountains (Mäkeläinen, 2017; Parkkinen, 2018b), which was appreciated by the Chinese panda experts (Lassi & Koski, 2018). Ranua, located in Lapland, has much longer and colder winters than Ähtäri.

The decision to give Ähtäri the pandas has triggered a lot of discussion in the months after the pandas landed in Helsinki. For instance, the publisher of the main Finnish tourism magazine Matkailulehti highlights that Ähtäri is too far from such domestic destinations that offer tourism services meeting the requirements of international tourists (Rautio, 2018). The main Swedish newspaper in Finland Hufvudstadsbladet questioned: "[I]f the pandas are so extremely popular, why are they placed in Ähtäri in the middle of the forest and far from all major cities in Finland"? (Westerlund, 2018). If international tourism flows were considered the pandas would have been placed in Helsinki Zoo (Korkeasaari) or in Ranua close to Rovaniemi, with 20 daily incoming flights during the winter season. The four hour drive by car from the capital area may be perceived as too exhausting for many families and international tourists (Rautio, 2018).

Moreover, the tourism policy of Finnish Government is implemented by Visit Finland, the national destination management organisation. For international markets, Visit Finland has divided Finland into four tourism regions with their own brands. The four regions are Lapland, Lakeland, Helsinki and the Archipelago. All four regions have their own strengths, a main season for tourism, identified potential customers and unique tourism supply (Roadmap, 2015). The municipality in focus for this study, Ähtäri, is located outside these main tourism regions. Especially in marketing activities focusing on the Swedish market Ähtäri is though often co-operating with Archipelago area due to the ferry connection between Vaasa (Finland) and Umeå (Sweden). However, there is a point in the criticism towards the deciding of Ähtäri Zoo to become the home of the pandas. Most international tourists visit either the capital area or Lapland.

Another fact is that Ähtäri and its surroundings have a very limited cluster of tourism services (Othman 2018; Rautio 2018b), which can be a challenge since visitors from Southern Finland need to stay overnight as Ähtäri is too far for a day trip. Moreover, even though the visitors are permitted to stay in the panda house the whole day (Rautio 2018a), the pandas are mainly eating and sleeping, resulting in that after one hour of watching most people would like to experience something else in the area nearby.

Local politics

Ähtäri municipality has a population of 6,000 (Ähtäri 2018), but as many other small municipalities in Finland it is struggling economically (Rautio 2018). In Finland, renting the pandas has not been criticised except for the high risk of the investments needed, for instance in the new panda house and the playground (Pääkirjoitus 2017; Sarpola 2018). In 2015 and 2016 Ähtäri was categorized as a municipality in crisis based on the Finnish municipality law (Karjalainen 2018; Neljä kuntaa täyttää 2018), which means that the finances of the municipality are struggling and the amount of debt is high. Thus, something had to be done and bringing in the pandas to the zoo was one attempt to get the finances balanced and consequently shedding the negative label of being a municipality in crisis (Rautio, 2018; Nuotio, 2017).

The idea was that the pandas will attract tourists who will spend money also on other services than entrance fees to the zoo. Moreover, some of the accommodation providers are planning extensions of the existing facilities, which will have multiplier effects in the local economy (Urpelainen, 2018). Thus, local people and political decision-makers hope for an improved image, financial benefits and jobs (Sarpola, 2018; Lumi ja Pyry pääsivät telmimään, 2018). Ähtäri is now branding itself as the city of pandas and the panda logo can be seen everywhere including the webpage of the municipality (Siltamäki, 2018a).

Figure 3. Finland divided into tourism regions (Visit Finland 2020)

The main target group is domestic tourists, but international marketing will also be initiated (Egutkina, 2018).

As discussed earlier, Ähtäri municipality owns 99 percent of the company running the zoo and local people, tourism experts and media have been worried about the risks with for the investments in the new panda house, at a cost of 8.2 million euros (Siltamäki, 2018a; Huhtanen, 2018). The company running the zoo needed to get the municipality as a granter of the loan (Karjalainen, 2018; Siltamäki, 2018a), which was accepted by the city council (Finland fick pandor, 2017). One local citizen got upset with this decision, asking if the local inhabitants could end up paying the house if something goes wrong. The issue was brought to the court which made the decision that the municipality should not have been allowed to give the grant for the loan (Huhtanen, 2018; Karjalainen, 2018). The fact that the company running the zoo did make a loss of one million euros in 2017 (Forsman, 2018) suggests that running a zoo is not that lucrative a business. Local politicians and civil servants are strongly believing that the panda project will be a success and put Ähtäri on the map in Northern Europe, if not globally.

The investment in the pandas should repay itself, which requires the number of visitors to Ähtäri Zoo to rise from 170,000 to 270,000 annually (Urpelainen, 2018; Huhtanen, 2018). This will not be easy since the tickets to the zoo are rather expansive, at 34 euros per person if the ticket is bought at the gate, and 39 euros if the visitors would like to see the entire zoo. The entry fee of Ähtäri zoo is much more expensive than entry fees at other zoos with pandas, such as Edinburgh Zoo at 21.5 euros, and 15 euros in Berlin. The reasons for the expensive tickets are the investments in a new building for the pandas and in the wellbeing of the animals in the zoo, as well as the regulation related to working hours and the taxation system in Finland (Rautio, 2018b). In addition to the income from ticket sales, the pandas are also used for additional products such as souvenirs, food and beverage, which will generate income for the companies in the area (Siltamäki, 2018a).

The manager of the Finnish research centre for Tourism (MIT), Antti Honkanen, stated in an interview that the hype about the pandas will not last forever (Sarpola, 2018). What is hoped, tough, is a romance between the pandas Lumi and Pyry resulting in some baby pandas (Koivula, 2018). A baby panda would entice visitors to come and see the pandas again.

8. Discussion

During spring **of** 2018, Ähtäri Zoo and its new inhabitants received a lot of media attention, and basically all Finnish main media channels (digital media, TV, radio) reported about the pandas. Pandas are perceived as cute animals and Finnish media reported about how the pandas settled in and what the pandas were doing when they were shown to the public for the first time. The politics behind the decision of giving the pandas to Finland and the especially Ähtäri

Zoo were also largely discussed. The main themes of this discussion are summarized in in Table 1.

International politics is an important aspect of China's panda diplomacy, both from the Chinese part but also from the perspective of the receiving country. Many articles analysed in this study discussed the Finnish decision to accept the pandas by acknowledging the fact that pandas are given to countries not criticising Chinese politics domestically and internationally. The decision to accept the loan of the pandas was not criticised though; media appear to be rather realistic with the fact that Finland as a small country is completely dependent on foreign trade and investments, and today's China is an important economic power.

Many articles also raised the issue of whether Ähtäri Zoo is the best home for the pandas, especially as they supposed to become a major tourism attraction. Ähtäri is a small municipality with the nearest bigger city, Tampere, a two-hour drive away. Of Finland's population of 5.5 million, 2 million live in the capital area. Ähtäri is a four-hour car ride away from the capital area. Geographically, Finland is a rather big country but has a small population. To keep most of the country inhabited, active regional policy supporting peripheral areas has been a key political strategy during the decades following the Second World War, when urbanisation (which started late) was intensified. According to a recent study, four out of five Finns would still like to see the whole of Finland inhabited and support that the state pay for it. (Tikkanen, 2018) A fact is that if a basic public service system is to exist in most parts of Finland over the next decades, some economic activity in the private sector must exist as well. In that sense, the decision to give the pandas a new home in Ähtäri is understandable.

The pandas did not end up in Ähtäri without politics, especially when the company running the zoo is almost fully owned by the municipality. Local politicians in Ähtäri were prepared to guarantee the loan needed by the company running the zoo, because the Chinese authorities would never have accepted the pandas to move in to a zoo without a new, modern panda house with a garden for the pandas to play in. It can be seen as a significant risk for a municipality already having a label of a municipality in crisis, but what would the alternatives be? In order to make money municipalities have to take risks. From the perspective of the municipality it is not only about whether the zoo will make a profit or not; visitors coming to see the pandas will spend money on many other services in the area. This will increase the turnover for many companies and some will be able to hire new employees. Municipalities cannot go bankrupt in Finland, but if the zoo runs into more economic problems the company owning it might have to go bankrupt. Thus, if the pandas attract a lot

Table 2. Policy making and the pandas in Ähtäri

Geographical level	What happened?	Reasons
International	Chinese panda policy Finnish decision at high political level to overlook the human rights situation and lack of democracy in China when accepting the "gift"	Softening the image Trade and industrial co-operation, China is Finland's 5th most important export market
National	Pandas were given to Ähtäri zoo by the Ministry of Agriculture and Forestry, even if no regional tourism cluster existed and the accessibility from the capital region is poor	Regional policy Good arguments for regional economic benefits by actors involved
Local	Many years of lobbying for the pandas, thinking outside the box to save the municipality Large risk by local city council when investing in new facilities despite the poor economic situation of the municipality	De-industrialisation and low profitability in farming Strong believe that the pandas will improve the image of Ähtäri and increase tourism income and employment possibilities

of visitors to the village, after a couple of years the zoo may look for some other animals that so far do not exist in Finland and hope for new wave of interest.

In the end, an important aspect in the panda discussion missing in Finnish media is how the pandas fit in with the existing tourism policy and strategies. So far, the focus has been on authentic experiences and on supporting the development of services in some destinations that are deemed to be attractive in international tourism markets as well. Based on the analysed articles it is rather evident that Ähtäri wanted to use pandas to increase tourism, but the whole project does not seem to be linked to the road map for how tourism in Finland is going to be developed through 2025, nor to any other thinking that has dominated Finnish tourism policy so far.

In all, this study identified some rather controversial issues related to political decision making related to receiving the panda couple to Finland and the municipality of Ähtäri. Most of the discussions in media took place when the decision to accept the pandas already was taken, instead of the politicians taking the discussion to the public during the lobbying phase. Short term financial gains influenced the policymaking rather than involving the public in a critical discussion about what implications the giant panda couple have for the image of Finland as a country and tourism destination. Pagilara et al. (2020) discuss the importance of citizens' trust in governments and highlight that trust is the final end holding communities together. To gain trust, the political decision making of governments at any geographical level must be characterised by transparency and core ethical values of the society.

9. Conclusion

The aim of the paper was to analyse the political decision making during the process of getting pandas to Finland and locating them in Ähtäri, a small municipality far from the big Finnish cities. During the analysis it became evident that most decisions were made based on economic reasoning; pandas enhance good relations with China resulting in more trade and investment opportunities for Finnish companies. The pandas were given to Ähtäri Zoo, since they prepared a plan stressing the local economic benefits of getting the new inhabitants to the zoo. Moreover, Ähtäri is a municipality in economic crisis and new ways of generating economic benefits are needed. Thus, the municipality was prepared to guarantee for the loan needed for all new investments by the company running the zoo. Visitors buy panda souvenirs, beer and chocolate in the local supermarket and eat panda ice cream in cafes. Ähtäri has become known as the panda city, at least in Finland.

The first months after that the panda house was opened the public in February 2018 appear to be a success. During these months the number of visitors to the zoo doubled (Siltamäki. 2018b). For instance, during the winter holiday weeks in March the panda house was visited by 500 persons a day (Egutkina. 2018).

The pandas are here to stay, most likely for the entire 15 years of the agreement. Even a Finnish WWF manager stated in an interview that the Ähtäri Zoo offers the best living conditions an animal can get in a zoo (Sarpola, 2018). Thus, as Mäkeläinen (2017) states:

"As the pandas are here- it is worth to visit Ähtäri zoo and with good mood- the rent paid for them is not dependent on the number of visitors rather the idling panda couple will cost the same one million euros a year anyway"

References

Anderlini, J. (2017, November 2). How the panda became China's diplomatic weapon of choice. *FT Magazine.* https://www.todayonline.com/world/how-panda-became-chinas-diplomatic-weapon-choice

Buchert, P. (2018, January 18). Mjuk päls döljer pandapolitiksens vassa klor. *HBL.* https://www.hbl.fi/artikel/mjuk-pals-doljer-pandapolitikens-vassa-klor/

Edgell, D. L & J., R., Swanson (2013). *Tourism Policy and Planning.* Routledge

Egutkina, A. (2018, March 16). Ähtärin Pandatalolla lupaava alku Pandoja on tultu katsomaan sankoin joukoin parjatusta lippuhinnasta huolimatta. *Iltalehti.* https://www.iltalehti.fi/kotimaa/201803152200814643_u0.shtml

Fennell, D.A. (2012). *Tourism and Animal Ethics.* Routledge

Fennell, D.A. (2015). The status of animal ethics research in tourism: A review of theory In Markwell, K. (Eds.), *Animals and tourism-Understanding diverse relationships* (pp. 27–40). Channel View Publications.

Finland (2020). *Encyclopeadia Britannica.* https://www.britannica.com/place/Finland

Finland fick pandor- mycket oklart kring deras kommande hem (2017, April 6). *HBL.* https://www.hbl.fi/artikel/finland-fick-pandor-mycket-oklart-kring-deras-blivande-hem/

Forsman, T. (2018, May 5). Ähtärin eläinpuisto vaikeuksissa. *MTV.* https://www.mtv.fi/uutiset/talous/artikkeli/ahtarin-elainpuisto-vaikeuksissa-pandat-rokottivat-kassaa-yhtio-pyytaa-kaupungilta-paaomitusta/6901994#gs.uO1eFVY

Frost, W. (2010). *Zoos and Tourism.* Channel View Publications

Hall, M. C. (2010). Politics and Tourism: Interdependency and Implications in Understanding Change, In Butler, R. & W., Suntikul (Ed.), *Tourism and Political Change* (pp. 7–18). Goodfellow Publishers

Hall, M. C. & Jenkins, J. (2004). Tourism and Public Policy, In Lew, A. A., Hall, M. C. & Williams, A. M. (Ed.), *A Companion to Tourism* (pp. 525–540). Blackwell

Harald, P. (2017, April 5). Kinas hundraårspresent till Finland: ett pandapar. *HBL.* https://www.hbl.fi/artikel/kinas-hundraarspresent-till-finland-ett-pandapar-2/

Helin, A. (2018). *Ähtärin eläinpuisto- Suomen paras eläintarha.* https://kerranelamassa.fi/kotimaan-matkailu/pohjanmaan-nahtavyydet/ahtarin-elainpuisto/

Hogenboom, M. (2013). *China's new phase of panda diplomacy.* https://www.bbc.com/news/science-environment-24161385. Retrieved 2018-06-05

Hughes, P. (2001). Animals, values and tourism- structural shifts in UK dolphin tourism provision. *Tourism Management,(22),* 321–329.

Huhtanen A. (2018, July 7). Ähtäri tarvitse 100 000 panda turistia. *Helsingin Sanomat.* https://www.hs.fi/kotimaa/art-2000005671898.html

Högmander J. & V. Vanhalakka (2018, January 18). Ähtärin pandat: Lumi ja Pyry ovat nyt Suomessa-Lue tästä kaikki, mitä sinun pitää pandoista tietää. *Aamulehti.* https://www.aamulehti.fi/uutiset/ahtarin-pandat-lumi-ja-pyry-ovat-nyt-suomessa-lue-tasta-kaikki-mita-sinun-pitaa-pandoista-tietaa-200675330/

Karjalainen, K. (2017 January 27). Ähtärin pandojen vuosihinta lähes miljoona euro. *Iltasanomat.* https://www.is.fi/kotimaa/art-2000005169433.html

Kasvun kärki matkailu 4.0- nopeaa kasvua ja työllisyyttä matkailusta (2017). *TEM.* https://tem.fi/matkailu-4.0-toimenpideohjelma

Koivula J. (2017, February 17). Ähtärin eläinpuiston pandatalo avasi ovensa: "Pandatalon rakentaminen oli ponnistus ja mittava investointi". *Maaseudun tulevaisuus.* https://www.maaseuduntulevaisuus.fi/ihmiset-kulttuuri/artikkeli-1.224607.

Kuntalehti Lumi ja Pyry pääsivät telmimään lumessa- pandatalo avautui vihdoin yleisölle. (2018, February 17). *Iltalehti.* https://www.iltalehti.fi/kotimaa/201802172200752619_u0.shtml.

Lassi, T. & Koski, S. (2018, January 17). Pandat ovat saapuneet Suomeen. *Iltalehti.* https://www.iltalehti.fi/kotimaa/201801172200674142_u0.shtml

Markwell, K. (Ed). (2015). *Animals and tourism-Understanding diverse relationships.* Channel View Publications

Mäkeläinen, M. (2017, April 5). Analyysi: Suomi pääsi Kiinan syleilyyn-Pandamaat eivät mesoa ihmisoikeuksista. *YLE.* https://yle.fi/uutiset/3-9547783

Neljä kuntaa täyttää kriisikuntakriteerit. (2017, June 5). *Kuntalehti.* https://kuntalehti.fi/uutiset/teuva-ahtari-jamijarvi-ja-hyrynsalmi-tayttavat-kriisikuntakriteerit/

Nuotio, T. (2017 August 10): Pelastavatko Pandat Ähtärin? *Aamulehti.* https://www.aamulehti.fi/kotimaa/kriisikunnaksi-vajonnut-ahtari-toivoo-pandapariskunnan-tuovan-kaupunkiin-miljoonia-euroja-200319389/

Othman, H. (2018, January 20). Kina tar bättre hand om sina pandor än oliktänkare. *Österbottenstidning.* https://www.osterbottenstidning.fi/Artikel/Visa/173981

Pagilara, F., Aria, M., Russo, L., Della Corte, V. & Nunko, R. (2020). Validating a theoretical model of citizens' trust in tourism development. *Socio-Economic Planning Sciences*. doi: https://doi.org/10.1016/j.seps.2020.100922

Parkkonen, T. (2017, April 6). Näkökulma: Pandat ovat söpömpiä kuin ihmisoikeudet. *Iltalehti*. https://www.iltalehti.fi/politiikka/201704052200097760_pi.shtml.

Parkkonen, T. (2018, January 18). *Ähtärin pandojen vuokrahinta yli 12 miljoona euroa*. *Iltalehti*. https://www.iltalehti.fi/kotimaa/201801182200678720_u0.shtml

Pastras, P. & Branwell, B. (2013). A Strategic-Relational Approach to Tourism Policy. *Annals of Tourism Research*, *(43)*, 390–414. http://dx.doi.org/10.1016/j.annals.2013.06.009

Pitkänen, K. (2017, April 6). Pandoilla pitäisi parantaa kiinalaisten eläinoikeuksia. *Aamulehti*. https://www.aamulehti.fi/maailma/pandoilla-pitaisi-parantaa-kiinan-elainoikeuksia-24396139/

Program of Government (2015). *Prime Minister's Office*. https://valtioneuvosto.fi/documents/10184/1427398/Ratkaisujen+Suomi_FI_YHDISTETTYnetti.pdf

Puoskari, B. (2014, December 19). Ministeriö haluaa pandat Ähtäriin- lappilaiset pettyivät. *YLE*. https://yle.fi/uutiset/3-7699290

Pääkirjoitus: Pandat pehmentävät liiaksi kuvaa Kiinasta. (2017, January 18). *Iltasanomat*. https://www.is.fi/paakirjoitus/art-2000005530251.html

Rautio, S. (2018a, January 18). Missä menee pandojen näkemisen kipuraja? *Iltalehti*. https://www.iltalehti.fi/kotimaa/201801222200686015_u0.shtml

Rautio S. (2018b, January 22*)*. *Rankkaa kritiikkiä pandojen sijoittamisesta Ähtäriin HS:ssä: " Suuri virhe"*. *Iltalehti*. https://www.iltalehti.fi/kotimaa/201801182200676904_u0.shtml

Ritchie, J.R.B & Crouch, G. I. (2003). *The Competitive Destination- A Sustainable Perspective*. CABI

Roadmap for Growth and Renewal in Finnish Tourism (2015*)*. *TEM*. https://tem.fi/en/roadmap-for-growth-and-renewal-in-finnish-tourism-in-2015-2025

Sarpola, L. (2018, January 18). Pörröiset turistihoukuttimet voivat kaatua veromaksajien harteille. *Iltalehti*. https://www.iltalehti.fi/kotimaa/201801152200669711_u0.shtml

Saunders, M., Lewis, P. & Thornhill, A. (2009). *Research Methods for Business Students (5th ed.)* Prentice Hall

Siltamäki, T. (2018a, July 7). Suurlähettiläät. *Helsingin Sanomat- Kuukausiliite.* pp. 20–27.

Siltamäki, T. (2018b. July 7) Ähtäri syntyi uudelleen hirvenä. Helsingin Sanomat- Kuukausiliite. p. 18.

Sojern, S. (2005). *What is Policy*? https://maytree.com/wp-content/uploads/544ENG.pdf

Suomen matkailun tulavaisuuden näkymät (2014). *TEM.* https://tem.fi/documents/1410877/2871099/Suomen+matkailun+tulevaisuuden+nakymat+17012014.pdf

Szczudlik, J. (2017). *Role of "Panda Diplomacy" in China's Foreign Policy.* https://www.pism.pl/publications/bulletin/no-83-1023

Teemoista sisältöä elämyksiin (2020). *Visit Finland.* https://www.businessfinland.fi/suomalaisille-asiakkaille/palvelut/matkailun-edistaminen/tuotekehitys-ja-teemat/tuotekehitys-ja-teemat-lyhyesti/

Tikkanen, S. (2018 July 24). Neljä viidestä haluaa, että koko Suomi pysyy asuttuna. *YLE.* https://yle.fi/uutiset/3-10313405

Urpelainen, A-K. (2018 January 13). Ähtäri odottaa pandaturisteja. *Kauppalehti.* https://www.kauppalehti.fi/uutiset/ahtari-odottaa-pandaturisteja-jos-60-000-kavijaa-vuositasolla-tulee-lisaa--niin-se-voisi-olla-viela-ihan-kannattavaa/KqAuNmvg

Veal, A. J. (2011). *Research Methods for Leisure & Tourism- A Practical Guide (4th ed.).* Pearson

Westerlund, T.(2018, January 18). Pandemi i vårt land. *HBL.* https://www.hbl.fi/artikel/pandemi-i-vart-land/

Wright, D. (2018). Cloning animals for tourism in the year 2070. *Futures.* Vol 95, 58–75

Ådahl, B. (2018, January 12). Pandat vuokrataan maille, joiden katsotaan suhtautuvan myötämielisesti Kiinaan- Eläimet saapuvat ensi viikolla Suomeen. *Aamulehti.* https://www.aamulehti.fi/uutiset/pandat-vuokrataan-maille-joiden-katsotaan-suhtautuvan-myotamielisesti-kiinaan-elaimet-saapuvat-ensi-viikolla-suomeen-200662424/

Ähtärin pandatalo avasi ovensa jonottavalle yleisölle. (2018, February 17). *Iltalehti* https://www.iltalehti.fi/kotimaa/201802172200752265_u0.shtml.

Ähtäri (2018). Ähtäri. Retrieved from http://www.ahtari.fi/kaupunki/

Ähtäri Zoo (2018). Ajankohtaista. Retrieved from https://www.ahtarizoo.fi/index.php/fi/

Äijälä, M., J.-C Carcia-Rosell & M, Haanpää. (2016). Kirjallisuuskatselmus. Eläimet osana matkailutoiminta. *Matkailututkimus* (12) (2), pp. 45–59.

Öberg, T. (2018 February 17). The Giant Panda Couple Frolic in the Snow for the First Time in Ähtäri Zoo. *Finland Today*. http://finlandtoday.fi/the-giant-panda-couple-frolic-in-the-snow-for-the-first-time-in-ahtari-zoo-watch-the-vdieo-and-pictures/

Anne Köchling

Tourist Relevance and Perception of UNESCO World Heritage Sites and National Parks in Germany: The Case Study of the "Wadden Sea"

1 Introduction

1.1 UNESCO World Heritage and National Parks in Germany

Unique cultural and natural heritage sites of outstanding value to humanity have been awarded world heritage site status by the United Nations Educational, Scientific and Cultural Organization (UNESCO) since 1972. Inclusion in the world heritage list is preceded by a lengthy application process in which the sites, or state parties in which the sites are located, demonstrate that at least one out of ten criteria required for inclusion is met and the prerequisites for protection and management of the site are fulfilled (UNESCO World Heritage Centre, 2019). As of 2020, 1,121 sites in 167 countries worldwide have been granted UNESCO world heritage status. There are currently 46 sites in Germany, including 43 cultural world heritage and three natural world heritage sites. The Schleswig-Holstein, Lower Saxony and Dutch sections of the Wadden Sea have been recognised as a UNESCO world heritage site since 2009. In 2011, the world heritage site area was expanded to include the Hamburg Wadden Sea and in 2014 the Danish area was added, so the entire Wadden Sea has been protected across borders since then (UNESCO World Heritage Centre, 2020a). The Wadden Sea is the largest coherent tidal flat landscape in the world and one of the last areas of European wilderness in which the natural environment has largely remained free of human influence (Gätje & Babinsky, 2008).

Prior to inclusion in the list of UNESCO world heritage sites, the German Wadden Sea region in Schleswig-Holstein had already been a national park since 1985, the Lower Saxony section followed in 1986 and the Hamburg section in 1990. In addition to the three sub-sections of the Wadden Sea, which are managed as individual national parks, 13 other German areas have been designated as national parks so far (Bundesamt für Naturschutz, 2020). The protection category "national park" in Germany is given through a legal declaration by the responsible political authorities of the states for coherent large protection areas. These areas have to be of special character in accordance with § 24 of the Federal Nature Conservation Act, fulfil the conditions of a nature reserve for a predominant part of their area and are largely in a condition that is not, or at most very slightly, impacted by humans (Bundesministerium der Justiz und für Verbraucherschutz, 2009).

1.2 Goals and Expectations for the Status UNESCO World Heritage Site and National Park

With the award, UNESCO pursues the goal of identifying sites that are especially worth being protected worldwide and preserving them for future generations (UNESCO World Heritage Centre, 2020b). Countries that have had their sites designated by UNESCO often follow further strategic objectives in addition to the aim of protection. They associate the internationally recognised brand "UNESCO world heritage" (Boyd & Timothy, 2006; Quack & Wachowiak, 2013; Raum, 2011; Ryan & Silvanto, 2011, 2009) with the chance of gaining prominence, optimising or sharpening their image and, concomitantly, giving (new) impetus to their tourism development (Quack & Wachowiak, 2013). In terms of tourism development, it is not only about achieving quantitative tourism growth. On the contrary, it is expected that the status associated with a qualitative award will help to gain the attention of target groups that are particularly interested in culture or nature, and thus to steer the visiting behaviour positively in the direction of protection (Park, 2014). Moreover, the clear focus on the target groups is also intended to counteract uncontrolled mass tourism (Boyd & Timothy, 2006).

At the same time, countries and sites hope to promote greater sensitivity to the protection of cultural or natural resources, as well as additional resources, i.e. through access to the world heritage fund, to third-party funding or to international cooperation. The award is also expected to promote stronger identification of the local population with their own heritage.[1] However, the positive expectations are often met with some scepticism from the local population, who, among other things, fear that the costs incurred and additional responsibilities may outweigh the benefits or they may be critical of an increase in the number of visitors (Quack & Wachowiak, 2013).

---

1 Rebanks Consulting Ltd., & Trends Business Research Ltd. (2009) provide one of 14 best practice case studies in their study which summarised an overview of various socio-economic effects of world heritage status.

For national parks as well, even though the ultimate goal is the protection of the undisturbed flow of natural processes in the natural dynamics of the macro chore, the opportunity to experience nature, education and recreational activities should be promoted (Bundesamt für Naturschutz, 2020). As tourism can help those responsible for national parks fulfil their mandate to promote such offers (Maschewski, 2008), the claim for protection in Germany is increasingly concerned with the added value of the national park brand for tourism development and its corresponding commercialization. In this way, "national parks have become trademarks for an intact natural landscape." (Maschewski, 2008, p. 15). In non-European countries, such as the USA, national parks were marketed much earlier for tourism (Hannemann & Job, 2003; Maschewski, 2008;).

As a general rule, with the awarding of national park status, the responsible parties on site also expect positive tourism development to follow (Maschewski, 2008). National parks are thereby considered to be special attractions that appeal to visitors who are interested in an extraordinary natural experience (Job, Woltering, & Harrer, 2009). It is also expected that national parks will be able to contribute to extending the season, promote cooperation among nature protection institutions and tourism industry within the national park region and improve the tourism infrastructure through nature experience offers (Bundesamt für Naturschutz, 2013).

1.3 *Existing Knowledge of the Actual Effects of the Designations*

The expectations of the impact of being awarded UNESCO world heritage site or national park status are therefore high. However, with regard to the actual impact of world heritage status on tourism development, science has so far failed to provide any consistent findings (Quack & Wachowiak, 2013). Ryan and Silvanto (2009) believe that the world heritage designation has a positive impact on the image of the holder and in turn can help to brand or re-brand a destination. Positive effects on revenue (Bundesministerium für Verkehr, Bau und Stadtentwicklung & Bundesamt für Bauwesen und Raumordnung, 2007; Wüpper, 2017), "awareness among all related parties" (Alakavuk & Helvacioğlu Kuyucu, 2009, p. 150; see also Rebanks Consulting Ltd., & Trends Business Research Ltd., 2009) or the destination selection (Li, Wu, & Cai, 2008; Wüpper, 2017) could be confirmed in various case studies. Other studies indicate that the world heritage designation has only a limited impact on visitor behaviour (Adie, 2017; Alakavuk & Helvacioğlu Kuyucu, 2009; King & Halpenny, 2014; Park, 2014) and the efficiency of the sites (Cuccia, Guccio & Rizzo, 2016).

Taking into consideration that the investigations are limited to individual world heritage sites and/or countries, the different results of the research are not surprising. Numerous other aspects, such as the political situation in the country, the infrastructure or the previous awareness of the country and the site, influence on travel behaviour in addition to the status. From an economic point of view, world heritage status appears to be more valuable to developing countries (Ryan and Silvanto, 2011) or to destinations that are still at the beginning of their destination life cycle (development and growth phases). Destinations that are in the phases of maturity and degeneration can obtain less benefit from the designation (Raum, 2011). Moreover, each world heritage site is unique and thus each has its own specific starting situation. The socio-economic benefits of the status are not self-fulfilling (Rebanks Consulting Ltd., & Trends Business Research Ltd., 2009). For example, the prerequisite for the creation of high awareness of previously lesser-known monuments is the professional marketing of the status (King & Halpenny, 2014; Steinecke, 2014).

The German national parks are generally located in peripheral, structurally weak regions. The designation can act as a unique selling point and help the parks position themselves as natural and environmentally friendly tourist destinations (Hannemann & Job, 2003; Job, Merlin, Metzler, Schamel, & Woltering, 2016), which in turn has positive effects on tourism development (Maschewski, 2008). The tourism and associated economic effects of the national park status are also extensively documented in the form of key figures (Job et al., 2016).

However, according to the existing literature, the visitor's national park affinity varies significantly, based on knowledge of the status when visiting and the motivation to visit being attributed to the status (Job et al., 2016). For example, the proportion of visitors to the Schleswig-Holstein Wadden Sea with a high level of national park affinity, at 17.1 %, is well below the nationwide average of 28.3 %. This is partly due to the fact that other main reasons for visiting the region lie inside the national park, which is located on the coast and thus a traditional bathing, spa and summer holiday destination. It is also believed that the additional awarding of world heritage status, besides all positive effects, could have caused some confusion among visitors with regard to the protection label (Job et al., 2016). Hannemann and Job (2003) also transfer the concept of the product life cycle to the national park destinations and determine that the national park status in traditional destinations such as the Wadden Sea, which is already quite advanced in its product or rather tourism area life cycle (Butler, 1980), can primarily contribute to reorientation. For example, through opening up the region to new visitor groups or, with the help of nature events, e. g. the brent goose days in spring, it can contribute to extending the season (Job et al., 2016).

Quack and Wachowiak (2013) conclude that brands for protected areas can influence destination choice, visitor motives, and travel behaviour, but the site's unique characteristics and experience value have more of an impact on destination choice than pure status. Taking the presented and partly controversial research findings into consideration, the study presented here contributes to the assessment of the relevance of the UNESCO world heritage and national park designations for the tourism development of destinations. It should help to answer the question to what extent the designations function similarly to brands.

1.4 Destination Branding as a Tool for the Positioning of World Heritage Sites and National Parks in Competition

Considering the growing competition between tourist destinations (Bieger & Beritelli, 2013; Eisenstein, 2018), the increasing interchangeability of destinations and the flood of information associated with the inflation of supply, the importance of the destinations' image has significantly increased with regard to travel decisions. As a result, the strategic goal of branding has become important in the management of destinations over the last 20 years (Eisenstein, 2018). Ritchie and Ritchie (1998) define a destination brand as

> "a name, symbol, logo, word mark or other graphic that both identifies and differentiates the destination; furthermore, it conveys the promise of a memorable travel experience that is uniquely associated with the destination; it also serves to consolidate and reinforce the recollection of pleasurable memories of the destination experience." (p. 103).

Identification and differentiation emphasise the classic functions of a brand (Aaker, 1991), but at the same time, as a special feature of the destination brand, they are based on the promise of a memorable, unique travel experience (Ritchie & Ritchie, 1998).

Tourism research now considers the implementation of a destination brand to be a (central) success factor in destination competition.[2] Destination branding, as a term, refers to all marketing activities that drive the branding goal and contribute to the creation of a destination image that positively influences destination choice (Blain, Levy, & Ritchie, 2005). However, the establishment of a

---

2 For an overview of the current state of scientific discussion see Eisenstein (2018).

destination brand first requires that world heritage sites and national parks can be autonomous destinations. Scherhag and Jäckel (2003) consider destinations in the context of brand management as

> *"geographical units (...) in which all accommodation, catering, entertainment/employment facilities necessary for a stay are available and which the guest has chosen as the destination for their trip."* (p. 98)

Thus, a destination with the designation of "UNESCO world heritage" or "national park" must include the adjacent region in the brand management to provide sufficient infrastructure for the visitors (Hannemann & Job, 2003). Nevertheless, the success of a destination is not dependent on the size, but on the uniqueness of the tourist offer (Laesser & Beritelli, 2013). This uniqueness can be confirmed by the awarding of the desgination to world heritage sites and national parks.

In contrast to other industries, however, some features make structuring of the brand more difficult in destination management. The product is created through the interaction of numerous individual service organisations. Therefore, the destination marketing organisation (DMO) responsible for the brand management can exercise little control and influence on the quality of the results and thus on the brand experience or redemption of the brand promise. This is made even more difficult by the fact that, due to the service nature of the product, the customer himself is involved as an external factor in the provision of services and he also has a higher, subjective purchasing risk (Eisenstein, 2018; Hankinson, 2012). Due to the perceived purchase risk, the image of destinations also has particularly high importance in the travel decision process (Pike, 2017). In order to reduce the complexity of selecting a destination, on the demand side, the only destinations that will be shortlisted for a stay in the destination area are those in which the associations, characteristics and competences connected to the client's image appear to be sufficient to fulfil the client's needs (Eisenstein, 2018).

The image of a destination brand consists of a client-perceived functional and a non-functional or emotional-symbolic brand benefit. The perceived functional benefit arises from the client's knowledge of the destination (e.g. with regard to the core elements of the offer, the landscape and the price-performance ratio). The perceived emotional-symbolic benefit is in turn determined by human traits or characteristics (brand personality) associated with the destination (Hosany, Ekinci, & Uysal, 2006). On the client side, this is an assessment of whether the brand seems appropriate for the fulfilment of sensual-aesthetic, hedonistic and social needs (e.g. recognition) (Burmann, Halaszovich, Schade,

& Hemmann, 2015; Eisenstein, 2018). Due to the increasing interchangeability of the purely factual offer on site (e. g. hotels, activities), as with other branded products (Aacker, Stahl, & Stöckle, 2015; Burmann et al. 2015), the emotional-symbolic benefits are of particular importance in the process of differentiation and brand positioning of travel destinations (Aacker et al., 2015; Burmann et al., 2015; Eisenstein, 2018; Ekinci & Hosany, 2006; Hosany et al., 2006).

Another consequence of the described complexity in the branding and management of destinations is that, despite the importance of external images, the market-oriented perspective, i.e. an image formation oriented exclusively to the tourism market, must be supplemented by an internal perspective in the destination branding. The internal perspective is used by the stakeholders to create a self-image of the internal target-groups for the development of the brand identity which is then mirrored by the external brand image (identity-oriented brand management) (Burmann et al., 2015; Eisenstein, 2018; Hankinson, 2012; Thilo, 2017). Taking the even greater management complexity, resulting from the objective of protecting world heritage sites and national parks, into consideration, the need for a holistic management approach involving all stakeholders is particularly important for world heritage sites and national parks wishing to build a brand. However, this assumes that the perception of the destination by external target groups (the brand image) is known by the stakeholders at the destination.

In the process of brand management, a brand positioning is defined as "how the brand differs from other competing brands and in which dimensions and characteristics it should be perceived as `better` than the others." (Aacker et al., 2015, p. 86). In addition to the knowledge of the external perception of one's own destination, it follows that a comparison/benchmarking to other destinations must be carried out in order to identify and use the attributes and unique selling points that have been positively perceived in the competition comparison (Laesser & Beritelli, 2013; Pike, 2009).

Due to the complex conditions mentioned, it is understandable that not every destination can become a brand (Eisenstein, 2018). Nevertheless, there are clear competitive advantages if a destination succeeds in establishing a destination brand. Brand owners have greater potential to differentiate themselves from the competition in the market, which leads to preference and loyalty on the side of the client and consequently to financial benefits for the stakeholders in the target area (Pike, 2009; Raum, 2011). In addition, the advantages for the traveller are the facilitation of choice in the decision-making process due to lower search costs, a risk reduction and possibly a gain in prestige (Pike, 2009). It is therefore not surprising that the Wadden Sea aspires to establish itself as an autonomous destination brand (Wadden Sea World Heritage Brand) in a cross-border context with the UNESCO designation (Common Wadden Sea Secretariat, 2014a) and has submitted a trilateral strategy for sustainable tourism in the Wadden Sea World Heritage destination (Common Wadden Sea Secretariat, 2014b).

Attractions such as the world heritage sites and national parks can be used for positioning purposes due to their proven uniqueness as unique selling points within a larger destination (Raum, 2011; Ryan & Silvanto, 2014). It is also possible, as in the case of the UNESCO World Heritage Site Wadden Sea, to be managed as an independent destination brand, including the adjacent region. In both cases, a situation analysis with regard to the demand-oriented perception (reputation/image) of the world heritage sites and national parks in comparison with the competition is indispensable as a basis for strategic decisions (Boyd & Timothy, 2006). So far, situation analyses with regard to the German world heritage sites and national parks were predominantly based on surveys on the economic effects or awareness and perception from the perspectives of the host communities or visitors at the respective site. This also applies to the National Park Schleswig-Holstein Wadden Sea or the UNESCO World Heritage Site Wadden Sea, which has performed a population survey since 2000 (LKN-SH/Nationalparkverwaltung, 2014; Nationalparkverwaltung Schleswig-Holsteinisches Wattenmeer, 2015) and also regularly participates in the guest survey Schleswig-Holstein (Nationalparkverwaltung Niedersächsisches Wattenmeer, & Nationalparkverwaltung Schleswig-Holsteinisches Wattenmeer, 2018). Supplementary to these valuable data, there has been a lack of information on the relevance of the UNESCO world heritage and national park designations for the making of travel decisions and the demand-oriented perception of German sites from the perspective of the entire German population. For the German world heritage sites and national parks, the present study makes an important contribution to this. By not surveying visitors, but a representative sample of the German resident population, insights can be obtained of the perception independent of a visit and the overall German market potential of the sites as travel destinations. The knowledge of the external brand image, as explained in Section 1.3, is essential in the course of branding and positioning efforts in order to obtain a realistic assessment of the market situation.

2 Methodology

2.1 Aims and Objectives of the Study

The present study was carried out by the inspektour GmbH with scientific support provided by the Institute for Management and Tourism (IMT) of the West Coast University of Applied Sciences in the summer of 2016. It pursued the goal of closing the mentioned gaps in the existing data on the touristic potential and perception of world heritage sites and national parks in Germany as well as potential branding effects of the designations. By doing this, it sought to provide useful results for tourism science and industry alike.

This study therefore examined

1. the general interest of German residents in a journey to UNESCO world heritage sites and national parks
2. the impact of the designations UNESCO world heritage site and national park on German residents' destination choice
3. the perception of all German UNESCO world heritage sites and national parks (as of summer 2016, i.e. 16 national parks and 41 UNESCO world heritage sites[3]) as well as four international reference destinations by German residents.

Previous literature, in this context, had predominantly provided individual case studies (Ryan & Silvanto, 2011). The advantage of the very extensive, and in this form unprecedented, data collection is the possibility of compiling a comparison of all German sites as well as four international best practice examples. In this way, the results create multiple analysis options and an assessment of the respective position in the competitive environment of the UNESCO world heritage sites and national parks. Since the formal bestowing of the designation requires all of the examined sites be subject to the same criteria, a comparison is legitimate despite all differences (Raum, 2011). The analysis of the perception of German sites also provides insights into which sites have the potential to position themselves as autonomous destination brands. As the Schleswig-Holstein Wadden Sea has both protection states, a comparison of the results for the two

---

3   Since then, the Caves and Ice Age Art in the Swabian Jura (2017), the archaeological border complex of Hedeby and the Danevirke and Naumburg Cathedral (2018), as well as the Erzgebirge/Krušnohoří Mining Region and the Water Management System of Augsburg (2019) have been added.

designations is also possible, which can also bring about interesting derivations for the purposes of marketing.

2.2 Methods

2.2.1 Study Design

The questionnaire design and data analysis were carried out in the project team comprised of members from the inspektour GmbH and the IMT. The Gesellschaft für Konsumforschung (GfK) was commissioned with the data collection. A sample, which was representative of the German-speaking population aged 14–74 living in Germany in private households (quota sample, total population equals 56,716 million people) was drawn for the survey within the framework of the GfK Consumer Panel. In June/July 2016, a total of 6,003 people completed an online survey administered by the GfK. All (as of 2016) 41 German world heritage sites and 16 national parks as well as four international benchmarks (the World Heritage Site of Memphis and its necropolis – the pyramid fields from Giza to Dahshur, the World Natural Heritage Site Great Barrier Reef, the National Park Hohe Tauern in Austria and the Swiss National Park) were included in the survey. Due to the high quantity of included sites, various subsamples were used, each of which was representative of the German-speaking population living in private households in the aforementioned age bracket. With regard to the questions on the perception of the investigated UNESCO world heritage sites and national parks, the number of people surveyed for each site was 1,000 respondents.

In addition to the questionnaire contents presented below, socio-demographic characteristics of the subjects were recorded so that extensive bi- and multivariate evaluations (e.g. in the form of target group analyses) are possible.

2.2.2 Survey Contents

The first part of the survey dealt with the general interest in visiting UNESCO world heritage sites and national parks. The interest in visiting both destination categories was measured by the proportion of the top two responses on a 4-point Likert-type scale with a range from "4=very interested" to "1=not at all interested". Secondly, the general relevance of the status UNESCO world heritage or national park for destination choice was quantified with a scale range from "4=very relevant" to "1=not at all relevant". Alternatively, the subjects could also choose "do not know" for both questions.

With regard to the measurement of the perceptions of German UNESCO world heritage sites and national parks, the model of the DestinationBrand study series (Eisenstein, Koch, Trimborn, & Müller, 2017), which has been used for destinations since 2009, was adapted for the study.

Based on this model, in the first step, the customer-oriented brand value of the UNESCO world heritage sites and national parks was measured with the help of the so-called "four-dimensional brand model" (Eisenstein et al., 2017; Eisenstein, Müller, & Koch, 2009; Koch, 2017). In the development of the model, similar multi-dimensional brand models from other industries served as examples (e.g. Perrey, Freundt, & Spillecke, 2015; Ramme, 2004).

The model is based on the assumption that, as a prerequisite for the formation of a brand image, awareness must be gained (1st stage) (Burmann et al., 2015; Eisenstein et al., 2017), in order to then generate a positive attitude towards the brand/likeability (2nd stage) and finally a willingness to buy (3rd stage) among the brand connoisseurs, which finally leads to an actual purchase in the last (4th) stage (Ramme, 2004). Following the Eisenstein et al. (2009) model for destination brands, this study investigated the awareness of UNESCO world heritage sites and national parks (1st stage) in both an unprompted (brand recall) as well as prompted (brand recognition), i. e. on a list with the requirement of awareness "even if only by name", manner (Aaker, 1991; Burmann et al., 2015). Secondly, the likeability (2nd stage) among brand connoisseurs (stage 1, awareness) was measured based on the proportion of top two responses on a 4-point Likert-type scale ranging from "4=very likeable" to "1=not at all likeable". Alternatively, respondents could choose to give no assessment ("do not know"). Third, the willingness to visit the sites was asked for. Here, the brand connoisseurs (stage 1, Awareness) were asked, which excursion or vacation forms (day trip from the place of residence or holiday location as well as a shorter or longer holiday trip) they would consider for the listed UNESCO world heritage sites or national parks. Multiple responses were possible (the 3rd stage). Finally, in the 4th stage, it was ascertained whether the respondent had already visited the UNESCO world heritage site or the national park in the past (in the last 3 years or more than three years ago). With the aid of the results of the individual process stages and the analysis of the transfer rates, i.e. the relationship between the stages, statements about the brand value of the world heritage sites and national parks can be made in the comparison of the competition.

In a second step, the image of the world heritage sites and national parks was examined in more detail. The importance of the emotional-symbolic benefit elements for the destination image was discussed earlier (see section 1.3). In order to test this area of benefit, which is particularly important for competitive differentiation, the subjects who had previously stated that they were aware of the UNESCO World heritage site/national parks ("if only by name", stage 1, awareness) were asked to what extent certain profile characteristics (19 in total) applied to the respective site. On the one hand, the attribution of certain traits such as "authentic/genuine" or "unique" was queried and on the other hand the attractiveness rating for five life phase types (e.g. "attractive for seniors (65 years and older)") was examined. The "characteristic attribution" was measured by means of the proportion of top two responses of a 5-point Likert-type scale ranging from "5 = fully applicable" to "1 = not at all applicable". Taking the ultimate goal of the world heritage sites and national parks – the protection of monuments and natural landscapes for future generations – into consideration, there was also a section containing an assessment of five sustainability aspects e.g. "environmentally friendly (ecologically sustainable)". All three pillars of sustainability (environmental, economic and social) (UNEP & UNWTO, 2005) were covered.

For the purpose of measuring the unique features of the images, the survey closed with open questions asking the respondents what came to mind when they hear the names of the world heritage sites and national parks under investigation (spontaneous associations).

3. Selected Findings

3.1. *Interest in the Sites and Relevance of the Designation in the Destination Choice*

More than 80 % of the German population stated that they were generally interested (top two values) in visiting a world heritage site or a national park, independent of a specific destination. The interest in visiting a national park was slightly higher (84 %, of which 45 % were "very interested") than in visiting a UNESCO world heritage site (81 %, of which 40 % were "very interested") (see Fig. 1).

A Kruskal-Wallis test showed significant differences in the interest to visit a UNESCO world heritage site in terms of the surveyed socio-demographic aspects of gender, children under 14 years in the household, age, education

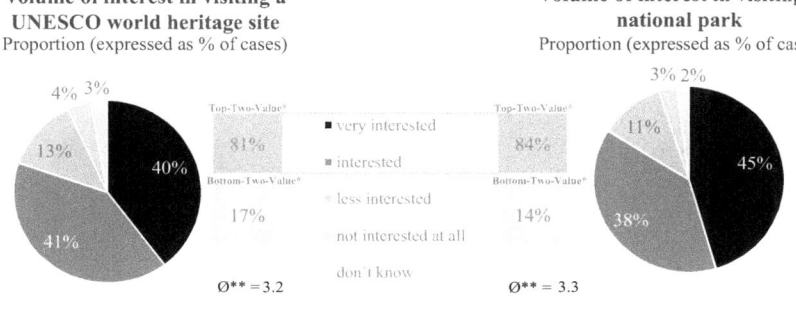

Fig. 1: Interest in visiting a UNESCO world heritage site and national park. Source: inspektour GmbH & FH Westküste, 2016

level, household net income and size of residence. The interest was above average among the respondents:

- between the ages of 65 and 74 (83.8 %)[4]
- with a university degree (86.3 %)
- with one child under 14 in the household (85.2 %) with a monthly net household income of over € 2,500 (84.1 %)
- who live in cities > 100,000 inhabitants (83.1 %)

With regard to the interest in visiting national parks, significant differences could be found in terms of the aspects of gender, age, education level and households with children under 14 years of age. Particularly standing out, due to the large degree of interest, were the interviewees

- aged 35–44 (85.9 %)
- with a university degree (88.2 %)
- with one (88.2 %) or two (89.1 %) children under the age of 14 in the household

---

4 It is also noticeable here that those under the age of 25 reported having a clearly below-average level of interest (74.9 %).

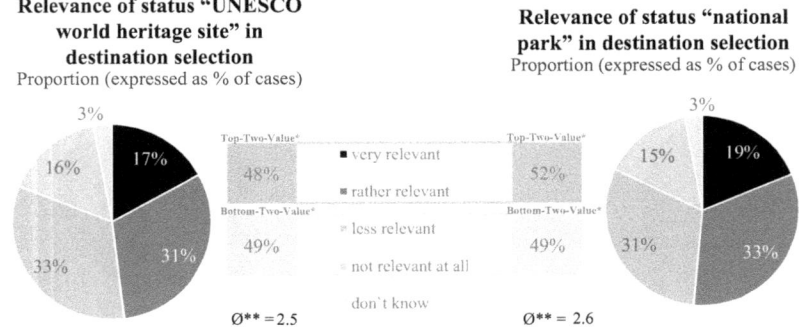

Fig. 2: *Relevance of the status UNESCO world heritage and national park in destination selection. Source: inspektour GmbH & FH Westküste, 2016*

When asked about the relevance of the UNESCO world heritage status in choosing the travel destination, almost half of respondents (48 %) said the status was "very" or "rather relevant". The relevance of the national park status was rated even higher at 52 % (see Fig. 2).

The Kruskall-Wallis test revealed significant differences in the relevance of the UNESCO world heritage status for destination selection for household size, children under the age of 14 in the household, age, educational attainment and household net income. The relevance assessment was particularly large in subjects

- aged 25–34 years (52.5 %)
- with a university degree (52.6 %)
- living in households with four or more people (53.0 %)
- with a monthly net household income over 2,500 EUR (53.0 %)

The relevance of the national park status for destination selection resulted in significant differences when taking the level of education, the size of the household and the number of children under the age of 14 in the household into consideration. Relevance attribution was high among respondents

- with a university degree (56.7 %)
- living in households with four or more people (56.7 %)
- with one (57.3 %) or two (59.6 %) children under the age of 14 in the household

The results indicate that the interest in visiting world heritage sites and national parks is high in the German source market. A very high relevance of the designations with regard to the destination choice could also be shown in the study. The interest in a visit and the relevance assessment for both designations were particularly high in the case of families with children and those with a high level of education. This corresponds with the educational mission of the national parks and the culturally demanding offer of world heritage sites and national parks.

3.2. *Perception with the Example of the UNESCO World Heritage Site Wadden Sea and the National Park Schleswig-Holstein Wadden Sea*

3.2.1. Brand Value

Overall, the study provides data on 43 UNESCO world heritage sites and 18 national parks, including four international benchmarks, as explained before. The findings for the UNESCO World Heritage Site Wadden Sea and the Schleswig-Holstein Wadden Sea National Park are presented here as an example.

In the first step, the degree of unprompted awareness was examined. With regard to the UNESCO world heritage sites, 7.7 % of respondents named the "Wadden Sea", which thereby had the second highest profile among all German world heritage sites. By far the highest unprompted level of awareness was held by the Cologne Cathedral with 18.8 % of respondents (see Fig. 3).

The unprompted awareness of the Schleswig-Holstein Wadden Sea National Park, on the other hand, was only 1.4 % of those surveyed, well behind other national parks. However, 12 % of respondents at this point named the Wadden Sea without further specification of the federal state. Therefore, these entries were not included in the ranking due to the lack of specification. The highest unprompted awareness of German national parks was achieved by the Bavarian Forest (20.4 % of respondents) (see Table 1).

This was followed by the measurement of the four-dimensional brand model (see chapter 2.2.2) where the first stage was concerned with the question of prompted awareness "even if only by name", i.e. with response specifications of

Top 10 responses: unprompted awareness national parks	% of cases	% of responses	number of responses
1 Bavarian Forest	20.4%	22.0%	218
2 The Harz Mountains	14.9%	16.1%	160
3 Eifel	12.5%	13.5%	134
3 The Black Forest	12.5%	13.5%	134
5 Saxon Switzerland	7.5%	8.1%	80
6 Müritz	6.0%	6.5%	64
7 Berchtesgaden	4.9%	5.3%	53
8 Jasmund	2.4%	2.6%	26
9 Hainich	2.3%	2.5%	25
10 Hunsrück-Hochwald	1.6%	1.7%	17
10 Lower Saxony Wadden Sea	1.6%	1.7%	17

▶ **Note:** The respondents on occasion did not literally name the travel destinations, but rather the information was put together from verbal cues and their corresponding individual answers.

Number of respondents: 1,071 / number of responses: 992 / Basis: all respondents with valid responses

Q: Which national parks in Germany do you know, even if only by name? (multiple response options)

Table 1: Unprompted awareness of German UNESCO world heritage sites. Source: inspektour GmbH & FH Westküste, 2016

Top 10 responses: unprompted awareness UNESCO world heritage sites	% of cases	% of responses	number of responses
1 Cologne Cathedral	18.8%	20.8%	201
2 *Wadden Sea*	*7.7%*	*8.6%*	*83*
3 Aachener Cathedral	7.0%	7.7%	75
4 Wartburg Castle	5.8%	6.4%	62
5 Speicherstadt and Kontorhaus district with Chilehaus Hamburg	5.7%	6.3%	61
6 Speyerer Cathedral	3.3%	3.6%	35
6 Hanseatic City of Lübeck	3.3%	3.6%	35
8 Town of Bamberg	2.9%	3.2%	31
9 Upper Middle Rhine Valley	2.6%	2.9%	28
10 Museum Island Berlin	2.3%	2.6%	25

▶ **Note:** The respondents on occasion did not literally name the travel destinations, but rather the information was put together from verbal cues and their corresponding individual answers.

Number of respondents: 1,071 / number of responses: 968 / Basis: all respondents with valid responses

Q: Which UNESCO World Heritage Sites in Germany do you know, even if only by name? (multiple response options)

Table 2: Unprompted awareness of German national parks. Source: inspektour GmbH & FH Westküste, 2016

the selected world heritage sites and national parks. Here, 74 % of respondents said they knew the UNESCO World Heritage Site Wadden Sea and 60 % the National Park Schleswig-Holstein Wadden Sea.

To measure the second stage of the four-dimensional brand model, subjects who were aware of the world heritage site or national park were asked how likeable the destination was. Measured by the top two values, the UNESCO World Heritage Wadden Sea achieved a likeability value of 68 %, the national park 55 %, among all respondents. In addition to the process stage values (based on all respondents), the transfer rates between the stages are also of interest, because these show the relationships between the individual process stage values. The transfer rate 1 indicates the proportion of the brand connoisseurs transferred to the next stage of being brand sympathisers. This value was 92 % for both of the Wadden Sea designations that were examined.

In the third stage, the willingness to visit was measured, divided into short (1–3 nights) and longer holiday trips (more than 3 nights) as well as day trips from the place of residence and from the holiday destination. About one-third (28 %) of the respondents stated they were willing to visit the UNESCO World Heritage Wadden Sea for a short visit, with 21 % indicating that they would visit the national park. The willingness to visit for a longer holiday is even greater at 40 % (UNESCO world heritage) and 31 % (national park). The willingness to visit for a day trip from the place of residence was naturally lower with regard to the total population in general due to the sometimes very long journeys that would need to be made (6 % UNESCO world heritage, 5 % national park). The willingness to visit for a day trip from the place of holidays reached lower values as well, i. e. 13 % (UNESCO) and 11 % (national park), due to the dependence on the location of the holiday destination. The transfer rates 2 indicate which proportion of the brand sympathisers are also willing to visit for the respective type of vacation. These transfer rates for the UNESCO World Heritage Wadden Sea were slightly higher than the values of the Schleswig-Holstein Wadden Sea National Park (see Fig. 3), based on holiday trips with overnight stays.

Finally, at the 4th stage of the four-dimensional brand analysis, respondents' visiting behaviour with regard to the world heritage sites and national parks in the past was queried. Almost half of the respondents (49 %) had visited the UNESCO World Heritage Wadden Sea before. Regarding the national park, the proportion of past visitors was 38 %.

Figure 3 summarises the presented results of the four-dimensional brand analysis for both designations in the form of a so-called brand funnel (see, e.g., Perrey et al., 2015). The willingness to visit is only presented here for the short (S) and longer (L) holiday trips for reasons of clarity. In addition, the findings

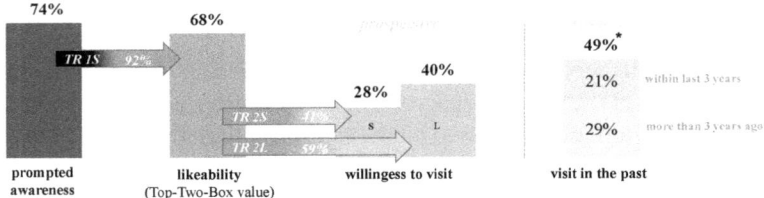

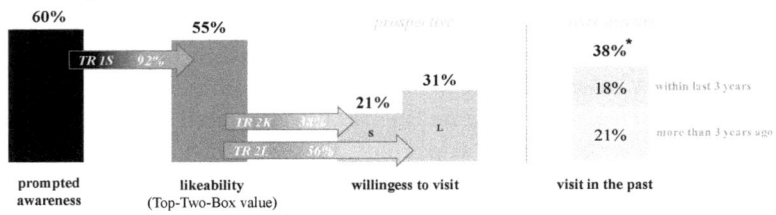

Fig. 3: *The four-dimensional brand analysis for UNESCO World Heritage Site Wadden Sea and Schleswig-Holstein-Wadden Sea National Park including the willingness to visit for future holidays (with min. 1 overnight stay). Source: inspektour GmbH & FH Westküste, 2016*

for each stage value were analysed according to socio-demographic criteria, and on this basis statements were made regarding distinctive features in brand values that were target group-specific.

The results were compared with all of the examined UNESCO world heritage sites and national parks in the form of a ranking and thereby the position in the competitive environment can be determined. The comparisons were made in different categories. As an example, a benchmark of the UNESCO World Heritage Wadden Sea is presented among all German UNESCO world heritage sites[5], as well as for the Schleswig-Holstein Wadden Sea National Park among all German national parks. The results are summarised in Table 3.

---

5   As mentioned before, there were 41 German world heritage sites at the point of time of this study. However, the transnational world heritage site in Stuttgart (architectural work of Le Corbusier) was awarded world heritage status only towards the end of the

Competitive analysis brand funnel	Category: UNESCO world heritage sites in Germany; number of competitors: 40			Category: national parks in Germany; number of competitors: 16		
	World Heritage Site Wadden Sea (N = 1,000)			National park Wadden Sea Schleswig-Holstein (N = 1,002)		
	own value in %	Ø of category	Ranking	own value in %	Ø of category	Ranking
prompted awareness as a travel destination (AW)	74%	32%	2.	60%	43%	4.
likeability (LI) (Top-Two-Value)	68%	27%	1.	55%	38%	3.
transfer rate 1S [= LI / AW]	92%	86%	7.	92%	89%	4.
willingness to visit for a day trip from place of residence(WDR)	6%	7%	18.	5%	5%	6.
transfer rate 2DHT [= WDR / LI]	9%	26%	40.	9%	13%	13.
willingness to visit for a day trip from place of holidays (WDH)	13%	9%	8.	11%	8%	4.
transfer rate 2DH [= WDH / LI]	19%	36%	40.	20%	22%	10.
willingness to visit for a longer holiday trip (WL)	40%	5%	1.	31%	17%	2.
transfer ratee 2L [= WL / LI]	59%	18%	1.	56%	42%	2.
willingness to visit for a short trip (WS)	28%	12%	6.	21%	18%	7.
transfer rate 2S [= WS / LI]	41%	43%	25.	38%	46%	16.
visit in the past	49%	19%	2.	38%	25%	4.

Table 3: Competitive comparison of four-dimensional brand analysis for UNESCO World Heritage Wadden Sea and Schleswig-Holstein-Wadden Sea National Park. Source: inspektour GmbH & FH Westküste, 2016

The UNESCO World Heritage Wadden Sea is ranked very highly in comparison. It is especially worth mentioning that in addition to the high level of prompted awareness (2nd place), it achieved the 1st place for both likeability and the willingness to visit for longer holiday trips. The Wadden Sea was ranked lower only in the willingness to visit in the context of day trips from the place of residence and the transfer rates for day trips overall. This could be explained by the decentralised situation of the region in northern Germany (with regard to day trips from the place of residence).

The Schleswig-Holstein Wadden Sea National Park was also in the lead in terms of process stage values in comparison to all 16 German national parks. In

survey phase in 2016. Therefore, it was not included in the competitive analysis, as it could not be ruled out that the associated reporting in the media could have had a significant temporary influence on the perception of this cultural world heritage site. Moreover, because of the subsequent admission, the number of recorded responses was significantly below 1,000 respondents.

addition, the second place in the willingness to visit for longer holiday trips was especially positive. The willingness to visit for short trips, however, was slightly lower coming in at 7th place (see Table 3).

It is clear that the UNESCO World Heritage Wadden Sea has shown very high brand value in the German source market. The degree of awareness, both prompted and unprompted, was very high. The brand also manages to generate likeability and the willingness to visit. The National Park Schleswig-Holstein Wadden Sea is also quite well positioned compared to all other national parks. In a comparison of the two Wadden Sea designations, the UNESCO world heritage scored better overall.

3.2.2. Profile Characteristics

In order to measure the emotional-symbolic aspects of the destination image (see sections 1.4 and 2.2), the subjects who were aware of the destination were asked to rate the extent to which certain characteristics applied to the UNESCO world heritage sites and national parks. The results were evaluated on the basis of all respondents, brand connoisseurs and visitors in the past and were also analysed in comparison of all German UNESCO world heritage sites and national parks. The following is a presentation based on the brand connoisseurs' responses, "even if only by name".

First, nine general character traits were tested. The highest approval values received by both Wadden Sea designations, also in the competitive comparison with all world heritage sites and national parks, were for the characteristics "nature-genuine" and "authentic/genuine". The characteristics "unique" and "sustainable" also received very high approval ratings. At the bottom of the rankings, the Wadden Sea was associated in both categories with the characteristics "culturally attractive" and "boring", whereby a low ranking in this specific characteristic should be viewed in a positive light. Concerning the aspect of "cultural attractiveness", the low ranking could be explained by the fact that the comparison was predominantly made with cultural world heritage sites as Germany only has three natural world heritage sites. Comparing the values for the two Wadden Sea designations, the approval values were relatively similar. The only noticeable difference was for the category "uniqueness", with more than three percentage points in favour of the natural world heritage designation (see Figure 4) Regardless of the designation, it is confirmed that the Wadden Sea has a fairly distinctive image in the German population.

In addition, respondents were asked to what extent the world heritage sites and national parks were attractive for selected life phase types. Again, the

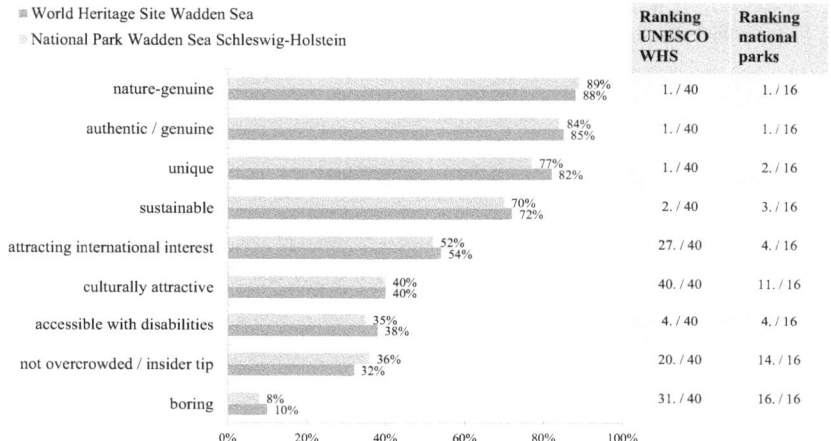

Figure 4: Agreement of brand connoisseurs with general characteristics: UNESCO World Heritage Wadden Sea and Schleswig-Holstein Wadden Sea National Park (proportion of top-two-box values and competitive ranking). Source: inspektour GmbH & FH Westküste, 2016

results for both Wadden Sea designations were very similar. The Wadden Sea was the most attractive from the perspective of elderly adults (31–64 years), followed by seniors (65 years and older) and families (with accompanying children under the age of 14). For the younger life phase types (young adults and adolescents), the levels of agreement were slightly lower, with the national park designation achieving somewhat higher values than the UNESCO world heritage designation. The Schleswig-Holstein Wadden Sea National Park was placed in the top ranks for all life phase types in a comparison between all German national parks – it was awarded a comparatively very high attractiveness regardless of the target group. When ranking the UNESCO world heritage sites, the UNESCO World Heritage Wadden Sea was also predominantly in the top ranks – but here there was a slightly more diverse picture. World heritage rankings were particularly high in terms of attractiveness for families (rank 1) and adolescents (rank 2), but slightly lower for seniors (rank 12) (see Fig. 5).

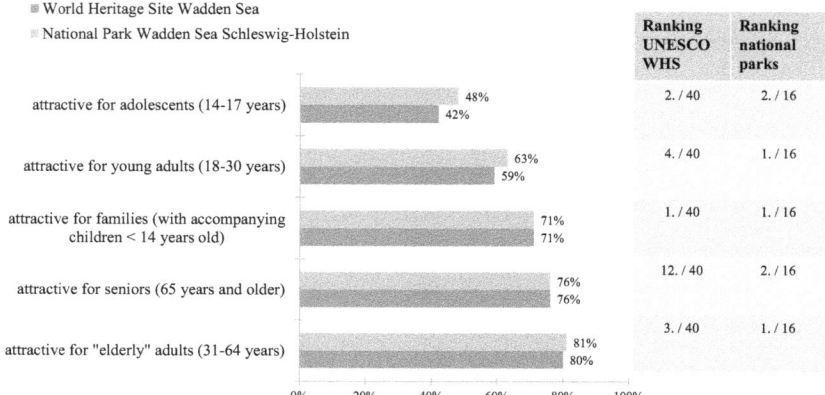

Number of respondents: at least 599 / Basis: respondents who are aware of the WHS / national park

Q: Just like people have certain qualities, the UNESCO world heritage sites and national parks can also be described by characteristics. For each of the following characteristics, please indicate to what extent you think they apply?

top-two-box values on a scale from "5 = fully applicable" to "1 = not at all applicable" (Proportion expressed as % of cases)

Fig. 5: *Agreement of brand connoisseurs with the suitability for life phase types: UNESCO World Heritage Wadden Sea and the Schleswig-Holstein Wadden Sea National Park (proportion of Top-Two-Box values and competitive ranking)*. Source: inspektour GmbH & FH Westküste, 2016

Finally, in the course of the profile characteristics, the agreement with certain sustainability aspects was measured. Differences of more than three percentage points between the two designations arose with regard to the attributes "economically efficient (economically sustainable)", where the national park received agreement values that were five percentage points higher, and "nature/environmentally friendly (ecologically sustainable)", where the UNESCO world heritage received a value five percentage points higher. Overall, the values for the attribute "of regional importance" were the highest for both designations, but the attribute "nature/environmentally friendly (ecologically sustainable)" also received very high agreement values. In turn, the attributes "socially just (socially sustainable)" and "economically efficient (economically sustainable)" received less agreement. However, looking at the detailed findings, these two characteristics were consistently given a very high proportion of "do not know" responses for all the UNESCO world heritage sites and national parks examined, which may be an indication that there were difficulties in assessing the sustainability aspect in accordance with the wording used in the survey.

The Case Study of the "Wadden Sea" 179

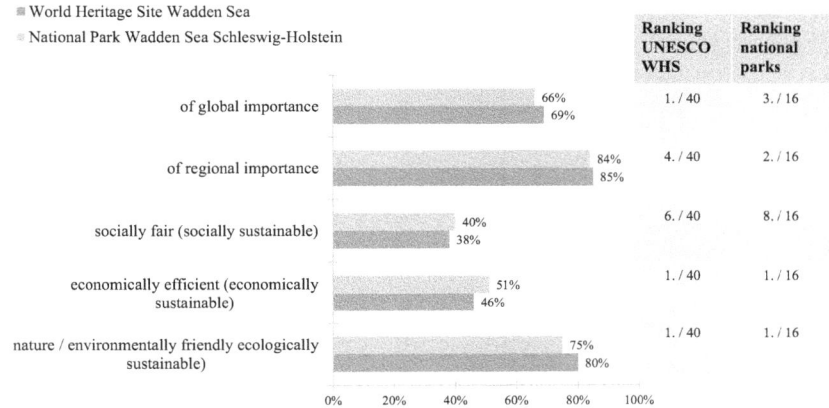

Fig. 6: *Agreement of brand connoisseurs with sustainability aspects: UNESCO World Heritage Wadden Sea and Schleswig-Holstein Wadden Sea National Park (proportion of top-two-box values and competitive ranking)*. Source: inspektour GmbH & FH Westküste, 2016

In the competitive comparison, both Wadden Sea designations were also in the top rankings when it came to sustainability aspects. Only with regard to the aspect of "socially just (socially sustainable)" does the Wadden Sea lose a few ranking places (see Fig. 6).

3.2.3. Spontaneous Associations

Moreover, to supplement the image measurement, the spontaneous associations of all UNESCO world heritage sites and national parks were measured. In comparing the two designations, World Heritage Site Wadden Sea and Schleswig-Holstein Wadden Sea National Park, more individual responses were given for the world heritage site. However, the content responses were very similar for both designations. Table 4 shows examples of the results of the top 10 individual responses for the World Heritage Site Wadden Sea. The associations of

Top 10 Spontaneous Associations World Heritage Site Wadden Sea (N=1,000)	% of cases	% of responses	number of responses
1. North Sea (-coast/ -landscape)	18.0%	11.1%	181
2. ((very) beautiful, long, interesting, North Sea-) (guided) walks across the mudflats at low tide	12.6%	7.7%	126
3. Ebb and flow / tides	8.0%	4.9%	80
4. ((very-) beautiful, unique, unaffected, rare, great, pure) nature / landscape / fascinating nature / art of nature	7.8%	4.8%	78
5. (wonderful at the) sea	5.8%	3.6%	58
6. Seal	5.5%	3.4%	55
7. (clear, a lot of, salty) water / (German) sea / waves / spume	4.8%	3.0%	48
7. Mud flats / (beautiful, interesting) wadden sea / biggest mud flats in the world	4.8%	3.0%	48
9. Generally positive assessments (e.g. (very) beautiful, unique, interesting, amazing, great)	4.2%	2.6%	42
10. lugworms	4.1%	2.5%	41

Table 4: Spontaneous associations with the UNESCO World Heritage Site Wadden Sea. Source: inspektour GmbH & FH Westküste, 2016

respondents related primarily to the particular coastal landscape and nature. In addition to aspects such as the location at the North Sea, the wildlife and possible activities, many emotional-symbolic aspects were also mentioned. This is evidenced by the numerous positive adjectives that were expressed within the framework of the associations.

The UNESCO World Heritage Site Wadden Sea and the Schleswig-Holstein Wadden Sea National Park therefore not only have high brand values in comparison with the competition (see section 3.2.1), but also a very clear brand image in the German population. In comparison with all German UNESCO world heritage sites and national parks, the Wadden Sea is positioned as a brand in the German source market.

Comparing the two designations shows that the World Heritage Site Wadden Sea has an even higher profile and brand value than the National Park Schleswig-Holstein Wadden Sea. Measured by the number of spontaneous associations, the world heritage image seems to be more recognised. On the other hand, the respective brand connoisseurs showed that the assigned profile characteristics are very similar. The Wadden Sea represents an authentic, sustainable and unique destination where the nature is paramount and can be experienced.

4. Conclusion, Critical Reflection and Outlook

In the German world heritage sites and national parks, the protection of the cultural or natural heritage goes hand in hand with the goal of sustainable tourism development. As a result, those responsible for the sites and parks are directly faced with the tasks of tourist destination management. There is a need for well-founded, scientifically-backed data to use as a basis for decision making, e.g. for branding, market positioning or target group analyses. By means of a population-representative survey in the German source market, the present study takes a perspective that has hitherto been neglected in research of UNESCO world heritage sites and national parks. Thus, the previously dominant guest perspective was supplemented by an overall population perspective for the purpose of creating visitor-independent potential estimates of the tourist relevance, prominence and image of the sites and the designations. Also unique is the very broad approach of the study, which included all German, as well as four international, UNESCO world heritage sites and national parks (as of 2016), thus enabling comprehensive competition analyses.

In the findings, great interest in visiting UNESCO world heritage sites and national parks could be established as well as a rather high relevance of the UNESCO world heritage and national park designations in the selection of travel destinations in the German source market. Thus, the study provides a scientific contribution to the discussion concerning the tourist relevance of the designations from the perspective of the entire German population (market potential). In addition, those responsible for tourism management in these protected areas receive valuable data from the detailed socio-demographic evaluations. Managers may compare these findings, together with additional market research information from the perspective of the guests on site (such as, in the case of the Wadden Sea, the guest survey Schleswig-Holstein conducted by the Institut für Tourismus- und Bäderforschung in Nordeuropa GmbH, 2018).

Furthermore, this study provides comprehensive data for the analysis of the perception of UNESCO world heritage sites and national parks in the German population, i.e. for customer-oriented brand value and for image in competitive comparisons. The results give insight into which sites have established themselves as autonomous destination brands.

The study shows that the Wadden Sea has high brand value and a clear profile in the German source market both under the UNESCO world heritage designation and as the Schleswig-Holstein Wadden Sea National Park. In a direct comparison of both Wadden Sea designations, the UNESCO world heritage designation scores slightly better than the national park designation. Hence, it can be confirmed for the German source market that the UNESCO World Heritage Site Wadden Sea is an autonomous destination brand. Building up the "Wadden Sea World Heritage" brand in the international context (Common Wadden Sea Secretariat, 2014a) seems to be the logical conclusion. In this context, especially taking the cross-border cooperation as well as the nature conservation objectives into consideration, reference should again be made to the approach of identity-based brand management (Burmann et al., 2015; Eisenstein, 2018; Hankinson, 2012; Thilo, 2017) and the associated necessity of involving all stakeholders in brand management. This study solely contributes data on the external image of the brands. This should be reflected in the self-image to be developed by all stakeholders within the destination.

The Wadden Sea in Schleswig-Holstein is a successful example of cooperation between the tourism industry and a national park authority. For a long time now, the regional tourism authorities have been using the positive image of the national park as a unique selling point in marketing. In addition, in the course of national park partnerships[6], cooperation between the National Park Office and tourist businesses and municipalities is carried out in order to jointly commit to the goal of sustainable tourism (Gätje and Babinsky, 2008). The world heritage status has further strengthened the desire for cooperation and is characterised by additional, supraregional and international projects (Nationalparkverwaltung Schleswig-Holsteinisches Wattenmeer, 2015). For example, a recently approved transnational inter-regional project is testing innovative strategies for small and medium-sized enterprises in the North Sea region to anchor the world heritage brand in the area, promote sustainable development and develop promotable products and services jointly with nature conservation stakeholders (Common Wadden Sea Secretariat, 2018).

With regard to the limitations of the study, it must be stated that the deliberately very broad research design (analysis of 61 sites and parks) had to be accompanied by a standardisation of the questions. An in-depth analysis, for example of the reasons for the relevance attribution of the designations or into how the subjects perceive the designations UNESCO world heritage or national park, therefore cannot be carried out and this question remains unanswered. Furthermore, regarding destination selection, the fact that the respondents were solely asked about the relevance of the designations and not about other

---

6   More information on the National Park Partnerships can be found at www.nationalpark-partner-sh.de (Nationalparkverwaltung Schleswig-Holsteinisches Wattenmeer, 2018).

aspects affecting destination selection limits the interpretation of the results at this point. When measuring the characteristic attribution of the sustainability aspects, the queried items should also be reconsidered in a new measure. The quite high number of "do not know" responses indicates that there were problems in understanding these questions.

Due to the fact that the positioning and image of a destination change only very slowly over time (Pike, 2009), it would be appropriate to repeat the study in about five years. In this way, a time comparison could provide insights into how the UNESCO world heritage sites and national parks have evolved.

References

Aacker, D. A., Stahl, F., & Stöckle, F. (2015). Marken erfolgreich gestalten. Die 20 wichtigsten Grundsätze der Markenführung. Wiesbaden: Springer Gabler.

Aaker, D. A. (1991). Managing brand equity. Capitalizing on the value of a brand name. New York, NY: Free Press.

Adie, B. A. (2017). Franchising our heritage: The UNESCO World Heritage brand. *Tourism Management Perspectives, 24*, 48–53. doi: 10.1016/j.tmp.2017.07.002.

Alakavuk, E. D., & Helvacioğlu Kuyucu, A. D. (2009). The role of UNESCO World Heritage List on cultural branding. Divrigi & Troy - two case studies from Turkey. In Pechlaner, H., Abfalter, D., & Lange, S. (Eds.), *Culture and creativity as location factors - looking beyond metropolitan areas* (pp. 139–154). Innsbruck: IUP - Innsbruck Univ. Press.

Bieger, T., & Beritelli, P. (2013). *Management von Destinationen.* 8. Auflage. München.

Blain, C.; Levy, S. E., & Ritchie, J. R. B. (2005). Destination branding: insights and practises from destination management organizations. *Journal of Travel Research, 43*, 328–338.

Boyd, S. W., & Timothy, D. J. (2006). Marketing issues and World heritage sites. In Leask, A. (Ed.), *Managing world heritage sites.* 1st edition (pp. 55–68). Amsterdam [u.a.]: Elsevier.

Bundesamt für Naturschutz (Ed.) (2013). *Weitere Nationalparke für Deutschland? Argumente und Hintergründe mit Blick auf die aktuelle Diskussion um die Ausweisung von Nationalparken in Deutschland.* Bonn. Retrieved from https://www.bfn.de/fileadmin/MDB/documents/themen/gebietsschutz/Nationalparke_Argumente-NLP10.pdf, checked on 03/02/2020.

Bundesamt für Naturschutz (Ed.) (2020). *Nationalparke*. Retrieved from https://www.bfn.de/themen/gebietsschutz-grossschutzgebiete/nationalparke.html, checked on 03/02/2020.

Bundesministerium der Justiz und für Verbraucherschutz (Ed.) (2009). „*Bundesnaturschutzgesetz vom 29. Juli 2009 (BGBl. I S. 2542), das zuletzt durch Artikel 1 des Gesetzes vom 15. September 2017 (BGBl. I S. 3434) geändert worden ist*". Retrieved from https://www.gesetze-im-internet.de/bnatschg_2009/__24.html, checked on 7/30/2018.

Bundesministerium für Verkehr, Bau und Stadtentwicklung (BMVBS), & Bundesamt für Bauwesen und Raumordnung (BBR) (Eds.) (2007). *Städtebaulicher Denkmalschutz und Tourismusentwicklung. unter besonderer Berücksichtigung der UNESCO - Welterbestädte*. Retrieved from https://d-nb.info/985201770/34, checked on 7/6/2018.

Burmann, C., Blinda, L., & Nischke, A. (2003). *Konzeptionelle Grundlagen des identitätsbasierten Markenmanagements. Arbeitspapier Nr. 1*. Retrieved from http://www.lim.uni-bremen.de/files/burmann/publikationen/LiM-AP-01-Identitaetsbasiertes-Markenmanagement.pdf, checked on 7/31/2018.

Burmann, C., Halaszovich, T., Schade, M., Hemmann, F. (2015). *Identitätsbasierte Markenführung. Grundlagen - Strategie - Umsetzung - Controllling*. 2. Auflage. Wiesbaden: Springer Gabler.

Butler, R. W. (1980). The concept of a tourist area cycle of evolution: Implications for management of resources. *The Canadian Geographer 24(1)*, 5–12.

Common Wadden Sea Secretariat (Ed.) (2014a). *Wadden Sea World Heritage Brand Paper*. Retrieved from http://www.waddensea-worldheritage.org/sites/default/files/downloads/wadden-sea_-wh-brand_paper-english.pdf, checked on 6/29/2018.

Common Wadden Sea Secretariat (Ed.) (2014b): *Sustainable Tourism in the Wadden Sea World Heritage Destination*. Retrieved from https://www.waddensea-worldheritage.org/sites/default/files/downloads/tourism-strategy-english-2014-12-22.pdf.

Common Wadden Sea Secretariat (Ed.) (2018). *Interreg North Sea Region: PROWAD LINK PROTECT & PROSPER: Benefits through linking sustainable growth with nature protection (PROWAD LINK)*. Unpublished project application.

Cuccia, T., Guccio, C., & Rizzo, I. (2016). The effects of UNESCO World Heritage List inscription on tourism destinations performance in Italian regions. *Economic Modelling 53*, 494–508. DOI: 10.1016/j.econmod.2015.10.049.

Eisenstein, Bernd (2018). Markenführung von Destinationen - Zwischen ökonomischem Nutzen, sozialer Konstruktion und Machbarkeit. *Zeitschrift für Tourismuswissenschaft 10 (1)*, 67–96.

Eisenstein, B., Koch, A., Trimborn, P., & Müller, S. (2017). Die DestinationBrand-Studienreihe. Basisinformationen zur Markenführung von Destinationen. In Eisenstein, B. (Ed.), *Marktforschung für Destinationen. Grundlagen - Instrumente – Praxisbeispiele* (pp. 267–283). Berlin: Erich Schmidt Verlag.

Eisenstein, B., Müller, S., & Koch, A. (2009). *Markenvierklang für Destinationsmarken - Modellentwicklung für die Markenstudie DestinationBrand*. Unpublished project report.

Ekinci, Y., & Hosany, S. (2006). Destination Personality: An Application of Brand Personality to Tourism Destinations. *Journal of Travel Research 45 (2)*, 127–139. DOI: 10.1177/0047287506291603.

Gätje, C., & Babinsky, M. (2008). Nationalpark Wattenmeer und Tourismus - Erfolgreiche Kooperation für Mensch und Natur. In Eilzer, C., Eisenstein, B., & Arlt, W. G. (Eds.), *National parks and tourism. Answers to a global question from the International Competence Network of Tourism Management (ICNT)* (pp. 59–82). München: Meidenbauer (Schriftenreihe des IMT, 3).

Hankinson, G. (2012). The measurement of brand orientation, its performance impact, and the role of leadership in the context of destination branding: An exploratory study. *Journal of Marketing Management 28 (7–8)*, 974–999. DOI: 10.1080/0267257X.2011.565727.

Hannemann, T., & Job, H. (2003). Destination „Deutsche Nationalparke" als touristische Marke. *Tourism Review 58 (Issue 2)*, 6–17.

Hosany, S., Ekinci, Y., Uysal, M. (2006). Destination image and destination personality: An application of branding theories to tourism places. *Journal of Business Research 59*, 638–642.

Institut für Tourismus- und Bäderforschung in Nordeuropa GmbH (NIT) (2018). *Gästebefragung (GB) Wattenmeer 2017. Repräsentative Befragung von Übernachtungsgästen an der Nordsee in Niedersachsen und in Schleswig-Holstein*. Unpublished project report.

Job, H., Merlin, C., Metzler, D., Schamel, J., & Woltering, M. (2016). *Regionalwirtschaftliche Effekte durch Naturtourismus in deutschen Nationalparken als Beitrag zum Integrativen Monitoring-Programm für Großschutzgebiete*. Bonn-Bad Godesberg: BFN-Skripten. Retrieved from https://www.bfn.de/fileadmin/BfN/service/Dokumente/skripten/Skript431.pdf, checked on 7/30/2018.

Job, H., Woltering, M., & Harrer, B. (2009). *Regionalökonomische Effekte des Tourismus in deutschen Nationalparken.* Bonn-Bad Godesberg: Bundesamt für Naturschutz (Naturschutz und biologische Vielfalt, 76).

King, L. M., Halpenny, E. A. (2014). Communicating the World Heritage brand: visitor awareness of UNESCO's World Heritage symbol and the implications for sites, stakeholders and sustainable management. *Journal of Sustainable Tourism 22 (5),* 768–786. DOI: 10.1080/09669582.2013.864660.

Koch, A. (2017). Markenbildung von Destinationen. Ein wichtiges Instrument im Angebotsdschungel. In Eisenstein, B., Weis, R., Reif, J., & Eilzer C. (Eds.): *Tourismusatlas Deutschland.* 1. Auflage (pp. 43–44). Konstanz: UVK Verlagsgesellschaft.

Laesser, C., & Beritelli, P. (2013). St. Gallen Consensus on Destination Management. *Journal of Destination Marketing & Management 2 (1),* 46–49. DOI: 10.1016/j.jdmm.2012.11.003.

Li, M., Wu, B.; & Cai, L. (2008). Tourism development of World heritage sites in China: A geographic perspective. *Tourism Management 29 (2),* 308–319. DOI: 10.1016/j.tourman.2007.03.013.

LKN-SH/Nationalparkverwaltung (Ed.) (2014): *Mehrwert Natur. Ein Gewinn für alle: Nationalpark, Weltnaturerbe und Tourismus.* Retrieved from https://www.nationalpark-wattenmeer.de/sites/default/files/media/pdf/broschuere-mehrwert-natur-web.pdf, checked on 8/1/2018.

Maschewski, A. (2008). Touristische Vermarktung Deutscher Nationalparke: Stand und Ausblick. In Eilzer, C., Eisenstein, B., & Arlt, W. G. (Eds.), *National parks and tourism. Answers to a global question from the International Competence Network of Tourism Management (ICNT)* (pp. 9–26). München: Meidenbauer (Schriftenreihe des IMT, 3).

Nationalparkverwaltung Niedersächsisches Wattenmeer, & Nationalparkverwaltung Schleswig-Holsteinisches Wattenmeer (Eds.) (2018). *2. Gästebefragung „Weltnaturerbe Wattenmeer und nachhaltiger Tourismus" 2017. Zusammenfassung der Ergebnisse der zweiten Gästebefragung im deutschen Wattenmeer im Jahr 2017.*

Nationalparkverwaltung Schleswig-Holsteinisches Wattenmeer (2015): Sozioökonomisches Monitoring (SÖM Watt) in der Nationalpark-Region. SÖM-Bericht 2015. Retrieved from https://www.nationalpark-wattenmeer.de/sites/default/files/media/pdf/soem-bericht-2015.pdf, checked on 8/1/2018.

Nationalparkverwaltung Schleswig-Holsteinisches Wattenmeer (2018). *Partner Nationalpark Wattenmeer.* Retrieved from https://nationalpark-partner-sh.de/, checked on 8/8/2018.

Park, H. Y. (2014). *Heritage tourism.* London: Routledge.

Perrey, J., Freundt, T., & Spillecke, D. (2015). *Power Brands. Measuring, Making, and Managing Brand Success.* 3rd Edition. Weinheim: WILEY-VCH.

Pike, S. (2009): Destination brand positions of a competitive set of near-home destinations. *Tourism Management 30 (6),* 857–866. DOI: 10.1016/j.tourman.2008.12.007.

Pike, Steven (2017): Destination positioning and temporality: Tracking relative strengths and weaknesses over time. In *Journal of Hospitality and Tourism Management* 31, pp. 126–133. DOI: 10.1016/j.jhtm.2016.11.005.

Quack, H.-D., & Wachowiak, H. (2013). Welterbe und Tourismus. Ausgewählte Forschungsergebnisse. In Quack, H.-D., & Klemm, K. (Eds.), *Kulturtourismus zu Beginn des 21. Jahrhunderts. Festschrift für Albrecht Steinecke. With assistance of Albrecht Steinecke* (pp. 279–295). München: Oldenbourg.

Ramme, I. (2004). *Marketing. Einführung mit Fallbeispielen, Aufgaben und Lösungen.* 2., überarb. Auflage. Stuttgart: Schäffer-Poeschel (Praxisnahes Wirtschaftsstudium).

Raum, S. (2011). *Der Markenwert touristischer Destinationen und seine Implikationen für das Destinationsmanagement aus tourismusgeographischer Perspektive. Herleitung und Überprüfung eines empirischen Messinstruments am Beispiel des UNESCO Weltkulturerbe Bamberg.* (Doctoral dissertation). Retrieved from http://opus.ub.uni-bayreuth.de/volltexte/2011/862/pdf/Diss_Sebastian_Bib.pdf.

Rebanks Consulting Ltd., & Trends Business Research Ltd. (2009). *World Heritage Status: Is there opportunity for economic gain? Research and analysis of the socio-economic impact potential of UNESCO World heritage site status.* Retrieved from http://icomos.fa.utl.pt/documentos/2009/WHSTheEconomicGainFinalReport.pdf, checked on 8/8/2018.

Ritchie, J. R. B., & Ritchie, R. J. B. (1998). The Branding of Tourism Destinations: Past Achievements and Future Challenges. In Keller, P. (Ed.), *Proceedings of the 1998 Annual Congress of the International Association of Scientific Experts in Tourism, Destination Marketing: Scopes and Limitations,* (pp. 89–116). Marrakech, Morocco. International Association of Scientific Experts in Tourism.

Ryan, J., & Silvanto, S. (2009). The World Heritage List: The making and management of a brand. *Place Brand Public Dipl 5 (4),* 290–300. DOI: 10.1057/pb.2009.21.

Ryan, J., & Silvanto, S. (2011). A brand for all the nations. The development of the world heritage brand in emerging markets. *Marketing intelligence & planning 29 (3),* 305–318.

Ryan, J., & Silvanto, S. (2014). A Study of the Key Strategic Drivers of the Use of the World heritage site Designation as a Destination Brand. *Journal of Travel & Tourism Marketing 31 (3)*, 327–343. DOI: 10.1080/10548408.2013.876956.

Scherhag, K., & Jäckel, M. (2003). *Destinationsmarken und ihre Bedeutung im touristischen Wettbewerb*. 1. Auflage (Doctoral dissertation). Köln: Eul.

Steinecke, A. (2014). *Internationaler Tourismus*. Konstanz: UTB.

Thilo, I. (2017). *Identitätsorientierte Markenführung im Tourismus*. Wiesbaden: Springer Gabler.

UNEP, & UNWTO (2005): *Making Tourism More Sustainable - A Guide for Policy Makers*. Retrieved from http://sdt.unwto.org/content/about-us-5, checked on 8/8/2018.

UNESCO World Heritage Centre (2019). *Operational Guidelines for the Implementation of the World Heritage Convention*. Retrieved from https://whc.unesco.org/document/178167, checked on 03/02/2020.

UNESCO World Heritage Centre (2020a). *World Heritage List*. Retrieved from https://whc.unesco.org/en/list/, checked on 03/02/2020.

UNESCO World Heritage Centre (2020b). *The World Heritage Convention*. Retrieved from http://whc.unesco.org/en/convention/, checked on 03/02/2020.

Wuepper, D. (2017). What is the value of world heritage status for a German national park? A choice experiment from Jasmund, 1 year after inscription. *Tourism Economics 23 (5)*, 1114–1123. DOI: 10.1177/1354816616655958.

Amber Knowsley, Tomas Pernecky and Jill Poulston

Motivations and Perceptions of Carbon Neutral Accommodation in New Zealand

1. Introduction

Carbon emission mitigation is of extreme interest to the global lodging industry, with the largest hotel groups now producing sustainability reports with specific emission reduction policies and implementation procedures (Bohdanowicz and Zientara, 2008; Jones, Hillier and Comfort, 2014; Legrand and Nielsen, 2017). The most common operational mitigation initiatives are recycling, energy efficient lighting, high efficiency appliances, sensors or timers to save electricity, and towel reuse options (Becken, 2013; Bruns-Smith *et al.*, 2015; Manaktola and Jauhari, 2007; Singh, Cranage, & Lee, 2014; Vernon *et al.*, 2003).

Describing mitigation efforts to stakeholders is an ongoing challenge, with no standardised industry guidelines, making comparisons difficult (Ricaurte, 2011), although various voluntary industry accreditation and certification platforms are available and provide some guidance. Some consumers feel that staying at 'green hotels' may compromise comfort (Barber and Deale, 2014), especially in an urban setting (Line and Hanks, 2016). Hotel operators therefore struggle to find ways to not only implement sustainability measures, but to do so without disrupting service quality (Rust and Oliver, 1994).

This exploratory case study seeks to understand the motivations and initiatives involved in curbing carbon emissions in accommodation and how these are viewed by customers. The overarching aim is to identify key issues and contribute towards developing a new theoretical framework. The following section reviews the literature on the most pertinent topics, and highlights that, to date, no research has examined this nexus. The methodology outlines the philosophical framework and tools employed for collecting and analysing data. The findings reveal poor communication of the hotels' motivations and initiatives, suggesting that corporate motivation and customer perspectives are likely to be mediated by enduring communication. It is posited that customers are likely to be more receptive of corporate motivations and intentions if these are communicated clearly, with rationales. A preliminary theoretical map is provided in the discussion with a framework for further research.

2. Literature review

The tourism industry is directly affected by the impacts of climate change, and is a "non-negligible contributor to climate change through GHG [Greenhouse Gas] emissions derived especially from the transport and accommodation of tourists" (Yang, 2010, p. 212). Projections forecast continued growth in international tourism with tourist arrivals worldwide expected to increase by 3.3 per cent a year from 2010 to 2030, reaching 1.8 billion by 2030 (UNWTO, 2015), however, a seven per cent growth was estimated for 2017 (UNWTO, 2018). Such growth will be unavoidably accompanied by proportional increases in associated emissions.

New Zealand is a developed country with a high-income (The World Bank, 2018), but nevertheless "potentially both highly vulnerable to climate change and highly economically dependent on tourism" (Scott *et al.*, 2012, p. 214). In the year to March 2017, tourism expenditure in New Zealand was NZD36 billion, with NZD14.5 billion from international visitors (Ministry of Business, Innovation and Employment, 2018). Tourism provides 8.4 per cent of employment (Ministry of Business, Innovation and Employment, 2018), and is New Zealand's leading export earner, overtaking dairy exports in 2016 (Tourism New Zealand, 2018a). Therefore, as climate directly affects the viability of destinations, tourist choices, and the industry as a whole (Hall *et al.*, 2015), New Zealand needs to be concerned with climate change.

Despite the clean, green image portrayed by Tourism New Zealand's "100 % Pure" campaign (Tourism New Zealand, 2018b), gross greenhouse gas emissions have risen by 23 % since 1990; New Zealand has the second highest level of emissions per GDP unit and the fifth highest per capita rating among members of the Organisation for Economic Cooperation and Development (OECD, 2017). The recent plan to introduce a Zero Carbon Act[1] – described as "a cornerstone of New Zealand's transition to low emissions, climate resilient future" (Zero Carbon Act, 2018, para 2) – signals the necessity to implement policy to curb carbon pollution. In this regard, it is also important to underscore the impacts of climate change within the lodging industry, due to its vulnerability and reliance on fixed assets (Su *et al.*, 2013).

Carbon emissions and sustainability in the lodging industry

Globally, the lodging industry is considered responsible for 21 per cent of all tourism carbon dioxide emissions (UNWTO, 2017), and forecast to account for a quarter of these emissions by 2035 (de Grosbois and Fennell, 2011). Hotels

---

1   "The Climate Change Response (Zero Carbon) Amendment Act passed in 2019"

account for around one per cent of total global GHG emissions (Gössling, 2011). Although this may seem insignificant, growth will increase contributions towards negative environmental impacts because of the enormous quantities of energy, raw materials, water and products used by accommodation providers (Michailidou et al., 2016).

Becken et al., (2001) observed a correlation between energy use and emissions per visitor night; as energy consumption rises, so do the emissions. Although hotels were identified as the largest energy consumers (110MJ equating to 7.9kg CO_2/ guest night) in the accommodation industry, potential exists to save energy across a range of services (Becken et al., 2001). According to Gössling (2011), hotels are high energy consumers because energy use is perceived to equate with comfort, exemplified by the use of 24-hour air conditioning, constant lighting, and heated swimming pools. He further notes that as a general rule, more luxurious accommodation uses more energy. This "high energy use is even, in some businesses, seen as a sign of quality and potency" (Gössling, 2011, p. xvii), connoting comfort and service quality to guests.

'Sustainability' has been defined as the interdependent and interlinked pillars of economic development, social equity, environmental protection and climate change (Brundtland, 1987; World Tourism Organisation and United Nations Environment Programme, 2008). Hotels adopting environmental sustainability as a core strategy are termed 'green hotels', described by Barber (2014) as "performing various environmentally friendly practices such as saving water and energy, using eco-friendly purchasing policies, and reducing emissions and waste disposals to protect the natural environmental and reduce operational costs" (p. 364).

Many large hotel groups are now committing to sustainability. Some claim that the lodging sector is in the midst of a sustainability awakening (Prairie, 2012), with environmental issues becoming recognised as increasingly important (Budeanu et al., 2016). Bohdanowicz and Zientara (2008) for example, note that a number of hotel companies (e.g, Accor International, Hilton Hotels, Club Mediterranean, Fairmont Hotels and Resorts, and Scandic) offer extensive information about their sustainable practices on their websites. All top ten global hotel chain brands (Heardable, 2012) provide information on their websites about their environmental sustainability, although only four publish formal sustainability reports (Jones et al., 2014). More recent research indicates that the largest hotel companies in the world are dedicating themselves to sustainability, with the five largest now publishing annual sustainability reports in addition to their traditional financial reports (Legrand and Nielsen, 2017).

Motivations for sustainability

Generally speaking, motivation in a business context emphasises profit maximisation (Okereke, 2007). However, recent research shows that accommodation businesses are "driven as much by a search for business efficiency gains as by a genuine concern for sustainability and the maintenance and enhancement of natural eco systems" (Jones *et al.*, 2014, p. 14). Corporate motivations for energy efficiency were compiled by Becken (2013), based on studies by el Dief and Font (2010) and Okereke (2007), and indicated that "apart from economic reasons, there are a number of reasons that motivate businesses towards energy efficiency, including stakeholder pressure, strategic proactivity, institutional dynamics, managerial ethics and organisational context" (p. 72). Based on the work of Levinas (1991), four motives for Corporate Social Responsibility (CSR) were identified by Roberts (2003). At the risk of oversimplification, these can be described as financial, ethical, social (i.e. being seen to be good), and individual, indicating the acceptance of responsibility for corporate behaviour at a personal level. This study focuses on ethical, financial and social reasons for sustainability, being readily recognised as corporate motives.

With regard to some of the challenges prevalent within the accommodation sector, Graci and Dodds (2008) identified that among the key obstacles in implementing environmental initiatives is a poor understanding of the associated benefits. They argue that businesses that 'go green' show improvement in the following areas: organisational performance, cost savings, employee loyalty, customer retention, risk management, public image and industry leadership. Others have similarly argued that, as sustainability emerges as a differentiator within the industry, hotels providing these "absolute necessities" (Rauch *et al.*, 2015, p. 101) sustainably, will build a strong competitive advantage.

The implementation of sustainable initiatives in hotels is mostly motivated by competitiveness or ecological responsibility (Bansal and Roth, 2000), but without a sole focus on carbon emission mitigation. Bansal and Roth suggest three primary motivations for ecological responsiveness: legitimation (i.e. compliance which is a financial influence, in terms of penalties for non-compliance), competitiveness (i.e. financial and social) and ecological responsibility (i.e. social and ethical), but without reference to Roberts' (2003) more comprehensive framework. The authors establish that legitimation (a financial motivation) is the strongest influence on change, and ecological responsibility the weakest. In their view, ecological responsibility has to do with genuine concern for "social obligations and values" and the environment (Bansal and Roth, 2000, p. 728). As industry is obliged to meet any legal obligations relating to carbon emissions, and ethical influences are weak motivators, the influence of consumer perspectives on competitiveness (i.e. financial and social motives) are likely to be strong motivators.

Applying Bansal and Roth's (2000) ecological responsiveness theory of corporate motivation suggests that competitiveness is the most prevalent motivator behind sustainability. As noted, luxury connotes consumption (Gössling, 2011), hence consumer perspectives are likely to be important influences on introducing environmental sustainability initiatives. Nevertheless, it is important to bear in mind that price and service quality can be more influential for consumers than sustainability, as found, for example, by Kasim (2004).

Perhaps not surprisingly, much of the research on customer perception of green hotels has concentrated on pricing (e.g., Chou and Chen, 2014; Chen & Tung, 2014; Han and Chan, 2013; Kuminoff, Zhang, and Rudi, 2010), but very little research has sought to examine more in-depth efforts of hotels to reduce carbon emissions and the way these are perceived by customers. No empirical evidence could be found for the motivations behind emission mitigation, pointing to a research area worthy of investigation, especially in terms of the relationship between curbing carbon emissions and customer responses to these initiatives. This exploratory case study therefore seeks to examine some of these nuances.

3. Methodology

This research adopts a pragmatist stance. As a theoretical position, pragmatism "privileges practice and method, over reflection and deliberative action" (Denzin & Lincoln, 2011, p. xii). The notion of focusing on practice and method influenced the choice of pragmatism, as the study aimed to produce "socially useful knowledge" (Feilzer, 2010, p. 6). In other words, rather than seeking 'absolute truths', the emphasis is on usefulness and utility– that is, the potential for theory to solve problems. Pragmatists consider that mixing approaches, methods and ideas helps "best frame, address, and provide tentative answers" (Johnson et al., 2007, p. 125).

Accordingly, a mixed methods research design was adopted to incorporate qualitative and quantitative methods within a single case study. This approach is particularly beneficial in tourism research as it helps to understand phenomena in their natural setting (Davies, 2003), allowing the verification and corroboration of information gathered through multiple methods.

A case study approach was selected to provide information about environmental sustainability and carbon neutrality in hotels that was not readily available from other sources (see Maxwell, 2010). The goal was to find a New Zealand accommodation provider committed to operating as carbon neutral. To source the case, the following keywords were entered into a Google search: "carbon neutral + accommodation + New Zealand", revealing just one potential participant: the Sudima Hotel Auckland Airport. The hotel's carbon neutral status was verified by searching for certified carbonNZero hotels in the Enviro-Mark Solutions database (https://www.enviro-mark.com/home), confirming that this was indeed the only carboNZero certified hotel and therefore an instrumental case. In addition, the Sudima Hotel Auckland Airport holds a Qualmark Silver award – a nationally recognised tourism accreditation awarded to companies for their dedication to creating an excellent experience for their visitors. It also has a Qualmark 4.5-star rating, denoting consistently high-quality levels of facilities and service (Qualmark, 2018).

Exploratory case studies are particularly suitable for exploring issues and problems where there is no prior research; they are thus seen as the "preliminary step of an overall causal or explanatory research design exploring a relatively new field of scientific investigation" (Mills, Durepos and Wiebe, 2010, p. 372). Their usefulness lies mainly in understanding new phenomena and generating ideas for further study, including the development of potential hypotheses and tentative theories. Hence, this research is inductive, and seen as a preliminary step towards building a new theoretical framework (see Eisenhardt, 1989). It also falls under the category of *instrumental case study research* (see Stake, 2005), whereby the goal is to gain insights into the interplay between corporate motivation and customer perception.

Data collection and analysis

In theory building case study research, it is common to use a combination of methods to gather data (Eisenhardt, 1989). Primary data in this study consisted of a face-to-face, open ended, semi-structured interview with the hotel's General Manager (GM) and a written questionnaire completed by the GM. The questionnaire comprised 30 questions on emission mitigation initiatives and motivations; its purpose was to further document and corroborate initiatives undertaken in the hotel. In addition, unobtrusive, nonreactive observation enabled the primary author to gather information during an overnight stay at the hotel, combating some of the methodological weaknesses of self-reporting (see Teddlie and Tashakkori, 2009) and providing additional validity (see Gillham,

2000). Field notes were made at the time and subsequently reviewed for contextual value towards building the case study as a whole.

Secondary documentation included analyses of publicly available documents such as three years' data from the hotel's carboNZero Summary of Certification, which provided historical information about the hotel, its emission sources, carbon dioxide equivalents for the period, and future reduction commitments. Additionally, 901 online TripAdvisor guest reviews of the hotel were analysed, specifically focusing on the emission mitigation initiatives most observed by guests – namely, the chilled beam air-conditioning and the absence of refrigerators in guest rooms. Reviews was grouped (positive, negative and neutral) and a summative content analysis applied to gain a deeper understanding of the hotel's overall service quality. Summative content analysis involves counting and comparisons, usually of keywords or content, followed by the interpretation of the underlying context (Hsieh and Shannon, 2005). Comparing information from multiple sources was utilised to enhance usefulness in context and obtain a "truer analysis" (Davies, 2003, p. 110).

This holistic contextualising strategy was employed to analyse the case study as a whole. In other words, the findings have been integrated in the analysis to provide a more holistic picture. This strategy interpreted the case as a complete text, looking for interconnections and disparities among different aspects (see Teddlie and Tashakkori, 2009). This contextualisation places an emphasis on the case itself, rather than breaking elements of it down into categories (Teddlie and Tashakkori, 2009), and was loosely modelled on Bohdanowicz and Zientara's (2008) case study of the Scandic Hotel Group. Validity was ensured through critical examination and 'disconfirming evidence', as in reality, evidence for themes diverge and include both positive and negative information (Creswell and Plano Clark, 2007). Triangulation (Creswell and Plano Clark, 2007; Teddlie and Tashakkori, 2009) allowed evidence to be provided separately from more than one source; as discrepancies were found, they were investigated further using various sources of evidence, a case study tool advocated by Gillham (2000).

Finally, it is worth noting that in theory-oriented case study research, sampling is theoretical rather than random (Eisenhardt, 1989). Accordingly, the Sudima Hotel Auckland Airport was selected, being the only carbon neutral hotel in New Zealand, and having the potential to offer valuable insights into the relationship between corporate motivation and customer perspectives. Many of the misunderstandings that surround case study research – including challenges to internal validity, generalisability and idiosyncraticity – have been addressed in other research texts; case studies can indeed provide deeper insights into emergent theories and sharpen "the limits of generalisability" (Eisenhardt, 1989, p. 544).

4. Results

Motivations for becoming carbon neutral

As a newly built hotel in 2011, the founding management decided that because of the initial investment in a sustainably focused building, they had a unique opportunity to implement emission mitigation initiatives throughout the hotel (see Leslie, 2015). Travellers can reduce their travel-based carbon footprints by staying in carbon neutral accommodation. With this in mind, management hoped to attract clientele interested in lowering their carbon footprints, thereby creating a unique selling point. Their commitment to environmental sustainability is reflected in the hotel's sustainability action points, published on its website (see Figure 1).

In an interview for AccomNews NZ, the hotel's owner explained that "going carboNZero ensured that our carbon footprint was neutralised, thereby giving us something concrete to relate to our guests" (see Leslie, 2015). The owner's support for Tourism New Zealand's "100 % Pure" campaign was a further motivation to take a sustainability approach, as noted in another interview:

> It seems to us that the combination of New Zealand's 100 % Pure tourism marketing campaign and high international visitor numbers means that our hotels should be walking the talk, and we are happy to be the first to demonstrate that carbon neutrality is possible in our industry (see Plaza, 2014).

Hotel emission mitigation initiatives

Recycling was an important aspect of day to day operations; paper and cardboard, glass, organics (food and garden waste) and plastics (but not aluminium) were all recycled. There was a back-of-house waste disposal and recycling process with different areas for dry waste, wet waste and cardboard, to maximise recycling efficiencies. The manager advised that waste management suppliers were selected based on their environmental processes as part of their carboNZero commitment:

> (The hotel) engages with people who handle our waste; it's not that we just let them go and pick it up and 'bye bye', but we look into these things. They are well monitored. (GM)

Additionally, according to the hotel's carboNZero Summary of Certification, the hotel's management was in the process of assessing how to treat organic waste, and considering an on-site organic waste disposal unit to reduce emissions.

A towel reuse option was offered on guestroom door hangers: "Hang me outside your door before midnight and your room will receive a light service

- Commitment to training and awareness of all team members in relation to Sudima Hotels' environmental and social improvement philosophy
- Ongoing development of environmentally-friendly procedures
- Consultation with all Sudima Hotels team members to foster ideas and awareness of the need to be a sustainable business
- Implementation of new policies at Sudima Hotels, as a result of consultation and research
- Engagement of suppliers to commit to sustainable practices and products
- Utilisation of local suppliers and contractors wherever possible
- Annual Green Globe benchmarking and Qualmark rating
- Raising of environmental and social issues at a local level (where applicable)
- Raising awareness among our hotel guests of our commitment to being sustainable
- Support of community initiatives aimed at sustainable use of the region's resources

Figure 1: The Sudima Hotel Auckland Airport's Active Management Plan ("Sudima Hotels", 2018)

the next day. Your choice to help reduce our environmental impact is greatly appreciated". Although this provided a towel reuse option, it was neither obviously located, nor specifically encouraging guests to save on laundry by reusing towels.

More than 2,000 LED (Light Emitting Diode) light bulbs were installed throughout the hotel. Management acknowledged that although LED bulbs were expensive, they were considered a worthwhile investment, lasting up to 50,000 hours (20 years of life for a single bulb), lowering waste and energy use. Another initiative was to purchase high Energy Star rated electrical items. The GM noted the "impact [savings] in operational costs, such as using energy saving appliances." The hotel supported a Switch Off policy to encourage employees and guests to turn off unused electrical items such as lights, and sensor lights [switches - remove] were fitted throughout the building, triggered by movement as someone entered or left a room. As the GM explained, "we practically don't have any light switches anymore" as everything was on a sensor, including lights in staff changing rooms and management offices. As the sensors turned off lights when there was no movement in a room, the only manual switches were electrical sockets. However, during observation it was noted that on arrival to a guest room, a standard lamp was turned on and operated by a manual wall switch, in contrast to the GM's claim that there were minimal light switches in operation. It is worth noting that with the installation of key card switches, main lights did automatically turn off when leaving the room.

Planford chilled beam air conditioning was installed throughout the hotel. This computer managed system uses hot and cold water to create different air temperatures, as well as allowing fresh air to circulate. According to the literature, these systems have a much lower operating volume than conventional hotel air conditioning units, and lower energy use (Kim *et al*., 2016). As commented by the hotel owners in the Indian Weekender (Nadkarni, 2011), despite the initial costs being higher, the long-term return made the investment worthwhile. The hotel's management also chose not to place refrigerators in the rooms. In the founding management's experience, this was justified because "…minibars are hardly used. They just consume a lot of electricity, need more resources to maintain and have over the years, just become another fixture in hotels." (Nadkarni, 2011, paragraph 15).

Further mitigation initiatives included the lack of paper compendia in guest rooms, which traditionally hold all relevant information about the hotel in a paper format. At the Sudima, this information was shown on in-room televisions instead. This saved money, lowered emissions and waste, and decreased labour by removing the need to provide physical updates of material. During the observation, however, it was found that the electronic compendium could not be accessed, despite sourcing a replacement remote control and assistance from staff. This technical issue was not explained, and was a negative aspect of the environmentally friendly compendium system.

Guest toiletries were provided in recyclable bottles made in New Zealand exclusively for Sudima Hotels and Resorts. The contents of the shampoo, body wash and lotion were biodegradable, GE (Genetically Engineered) free and not tested on animals. However, investigation revealed that the manufacturers were not certified with carboNZero, contrary to the hotel's survey data, which indicated they used a particular guest toiletries supplier because they were certified by carboNZero. Although the hotel's management was asked to comment on this discrepancy, no reply was received during the study period.

Management engaged in further emission reduction initiatives such as the installation of low flow shower heads and taps, which were difficult to verify during the observation as they look identical to conventional showerheads and taps. Survey data indicated that toilets had half flush options, which was verified during the observation. Although sourcing of local produce was not indicated in the survey data, some produce for the hotel's restaurant came from an on-site organic vegetable and herb garden. This lowered their food mile emissions, as well as providing guests with fresh organic produce. Furthermore, the GM indicated that the organisation was interested in measuring and off-setting guests' emissions, but had not yet undertaken it.

Additional environmental sustainability features were implemented that were not directly regarded as emission mitigation initiatives, although they were part of the hotel's overall environmental sustainability commitment. A non-chemical, indoor, heated pool was cleaned using an Enviro-swim ES–3 Electronic Oxidisation (ORP generator/Ionisation) system. This system uses copper and silver to sanitise the water, creating a low maintenance system which positively benefits the environment and guest health, and lowers costs (Watertech Plus, 2016). Cleaning products by American based Ecolab were used throughout the hotel. The main cleaning product was the Neutral Disinfectant Cleaner, a "multi-purpose, neutral pH, germicidal detergent" (Ecolab, 2016). The hotel has a north-south orientation to maximise the sun's energy in the southern hemisphere, and the gardens were planted with New Zealand native trees and plants, reducing the need for soil additives and artificial watering systems. The building had a 25,000 litre tank for harvesting rain water for non-potable use.

The most recent initiative was to install electric car charge stations in the hotel's car park. Future targets included a feasibility study for the installation of solar panels, a gas and waste audit, installation of a heat exchanger for the domestic hotel water supply and an organic waste treatment system. Finally, the hotel offset the remainder of its unavoidable emissions by purchasing carbon credits through carboNZero. The Sudima Hotel Auckland Airport undertook many emission mitigation and environmental sustainability initiatives beyond those required to maintain its carboNZero status, whilst still maintaining a high level of service quality, as observed during the overnight visit and through the retention of its Qualmark 4.5-star and Qualmark Silver status.

Corporate management motivations

The survey completed by the GM highlighted whether the hotel management considered 'financial savings' (competitiveness) or 'emission reductions' (ecological responsibility) as motivations for each mitigation initiative. Ecological responsibility was the corporate motivation behind the design and build phase of the hotel, an undertaking strongly recommended by the Project Manager and GM when developing a carbon neutral establishment. However, for the specific initiatives investigated, only recycling was reported to be driven by ecological responsibility. Although the installation of energy efficient lighting was noted by the GM to "...really cut operational costs", it appeared that reasons for installing this were equally motivated by competitiveness and ecological responsibility. The motivations for selecting high Energy Star rated appliances for the hotel were to lower energy costs and cost of purchase (competitiveness), over lowering carbon emissions (ecological responsibility). This suggests that competitiveness also motivated this initiative. Reasons for implementing an organisation-wide Switch Off policy were reported by the hotel to be equally for competitiveness and ecological responsibility, and the towel reuse option also appeared to be equally motivated by both competitiveness and ecological responsibility. Results overall indicated that purchases of highly efficient energy use items were driven mainly by competitiveness, and recycling completely by ecological responsibility, however, the other three initiatives were equally driven by both motivators. All data on motivations were derived from the hotel's GM.

Guest awareness of mitigation initiatives

Data were also collected on customers' awareness of the hotel's environmental undertakings through manual data mining of the hotel's TripAdvisor site (Sudima Auckland Airport Hotel, 2018). The two most uncommon and conspicuous emission mitigation initiatives were investigated – chilled beam air conditioning and removal of refrigerators from guest rooms. Although temperature is subjective for guests, results showed that of the 33 reviews mentioning air conditioning, 22 were negative (66.6 per cent). Customer reviews illustrated dissatisfaction with this:

- The air conditioner control on the wall is very ineffective and slow in response.
- There is a [sic] air conditioning system called "Chilled Beam Air conditioning", this is supposed to be very enviro friendly compared to traditional air con...not that impressive being very slow to react to any temperature change. In fact, my room was very cold (winter here in Auckland) and, we just could not get the room warm.
- The enviro friendly chilled water system aircon is very slow to do anything, however eventually does the job (read: hours later). Certainly, it's very quiet and that's a good thing. Many clients such as us however would prefer more immediate control of the temperature in the room.

These reviews showed dissatisfaction with both the responsiveness and effectiveness of the air conditioning. This service quality problem was directly related to a key emission mitigation initiative, and needed addressing by the hotel's management, but due to timing, this issue was not discussed with the GM. TripAdvisor data also showed that the lack of refrigeration in guest rooms was unpopular, and the most commonly mentioned issue in reviews. Of 901 reviews, 67 (7.4 %) mentioned this particular initiative negatively:

- I would have expected a small fridge in the room, unfortunately ours didn't have one.
- Room was generous in size and well-appointed although no mini fridge.
- The room was of the quality I expect of Sudima but not having a fridge in the room was quite a shock. There was certainly room to put one in so I'm not sure why they haven't.

In reaction to comments, an Assistant Manager's response on TripAdvisor noted that:

- Not having fridge in the room has been one of our efforts in support to [sic] the environment which has succeeded us [sic] in achieving New Zealand's first carcertification.

Although some reviewers seemed not to know why they had no refrigerator in their room, a clearly displayed brochure (on recycled card) provided an explanation: "our contemporary, fridge-less rooms have been estimated to reduce our carbon footprint by 236 trees per year. Five years = 1,180 trees". Some guests (n= 4) were however, aware of the environmental reasons for having no refrigerator, and while noting the lack of refrigeration, also acknowledged the environmental rationale for the decision, albeit with some cynicism:

- You MUST book a fridge as all rooms have had their refrigerators removed to be more 'green'.
- The hotel has no mini-bar, and they proudly proclaim this is saving them in power bills (I suppose they are right).
- Room fantastic albeit no fridge which is described in hotel info as part of a 'green' strategy to reduce CFC's.
- There are no mini-bar fridges in the hotel rooms to reduce carbon emissions.

Others noted that their room did have a refrigerator (n=2), that it was fine there was no refrigerator (n=3), or that they were happy there was no refrigerator (n=9). One commented:

- No bar fridge in the room (on enviro grounds) and I like that because the noise of them drives me spare.

Although reviews revealed some positive guest attitudes to this emission mitigation initiative, of 85 reviewers who mentioned refrigeration in any context, 78.8 per cent (*n*= 67 reviews) reviewed the lack of refrigerators negatively.

Despite the hotel owners' strong conviction that "travellers – whether for business or pleasure – are factoring in sustainability when they decide which hotels to patronize [sic]" (Jhunjhnuwala as cited in Leslie, 2015, paragraph 15), few guests mentioned sustainability in their reviews on TripAdvisor between 2011 and 2016. The hotel's reviews were also examined using the keywords of green, enviro, environ, environmental, environmentally, eco, sustainable, sustainability, emissions/carbon emissions, carbon/carbon neutral/carbon friendly/carbon footprint/footprint. Summative content analysis revealed the scarcity of guest reviews mentioning any aspect of sustainability, as presented in Table 1.

Table 1: Summative analysis of The Sudima Hotel Auckland Airport TripAdvisor guest reviews (2011 – 2016)

Key Search Word	Frequency	Example
Green	5	All rooms have had their refrigerators removed to be more 'green'. Pride yourself on being green.
Enviro/environment /environmental/environmentally	4	The hotel makes a big thing about being green and enviro friendly.
Eco/eco-friendly	3	A nice balance of value for money, a little luxury, eco-friendly. Eco friendly hotel. . .a beautifully appointed hotel
Sustainable/sustainability	0	
Carbon/carbon neutral/carbon friendly/carbon footprint/ footprint	4	The hotel is new and the rooms are pleasant and comfortable. The hotel has been built to be environmentally friendly and leave a small carbon footprint. It is a modern, clean hotel with contemporary decor and is leading the way environmentally by reducing its carbon footprint.
Emissions/carbon emissions	1	There are no mini bar fridges in the hotel rooms to reduce carbon emissions.

Data in Table 1 highlight the scarcity of guest reviews mentioning any aspect of sustainability, with less than one per cent for each key word(s). While this may be because the forum is not conducive to discussions or reviews regarding sustainability, is important to note that disappointed guests may be strongly motivated to leave a negative review (Radojevic *et al.*, 2015). It could also be inferred that because the hotel succeeds in providing guests with service quality and amenities sustainably without any deterioration of service (see Kirk, 1995), guests simply did not feel the need to provide online reviews of these services.

No reviews mentioned the hotel's carbon neutral or carboNZero accreditation status. Although the hotel's owners believed eco-awareness was an important part of decision-making for guests – noted for example, in an interview for AccomNews NZ (Leslie, 2015) - this was not strongly reflected in their comments. Only one review noted the hotel's carbon emission scheme, with most reviews commenting on the accommodation being 'green' and 'environment friendly'.

Environmental sustainability as a competitive advantage

An analysis of the Sudima's corporate website did not reveal any mention of its carboNZero certification, and its "Environmental Sustainability" page made no mention of its carbon neutral rating amongst the numerous other sustainable initiatives discussed (see Figure 1). Furthermore, although the "Environmental Sustainability" information mentioned its dedication to Qualmark and Green Globe ratings, only the Qualmark Silver certification was displayed on the company's homepage with a discreet logo. Reasons for the lack of carboNZero representation on the website were sought from the GM, who explained that the hotel was owned by a group, and head office had not updated the website to reflect the hotel's carbon neutral status. However, two years after the interview, the website had still not been amended. Additionally, the hotel does not have a "TripAdvisor GreenLeader Program" award (see https://www.tripadvisor.co.nz/GreenLeaders), which represents another missed opportunity for attracting 'green travellers' due to its internationally recognised status.

5. Discussion and conclusions

The Sudima Hotel Auckland Airport provided valuable operational and motivation insights into the first carbon neutral certified hotel in New Zealand. Previous studies found that the five most widely adopted mitigation initiatives in the accommodation industry are recycling, energy efficient light bulbs, other energy reduction methods (such as installing high starred Energy Star appliances), sensors or timers to save electricity and a towel reuse option (Becken, 2013; Bruns-Smith *et al.*, 2015; Manaktola and Jauhari, 2007; Singh, Cranage, & Lee, 2014; Vernon *et al.*, 2003). Becken (2013, p. 85) referred to these initiatives as "low hanging fruit" indicating they are relatively easy to implement. In this case study, the hotel had implemented all of them. Moreover, the hotel's management not only fulfilled the mitigation requirements set by carboNZero, but also implemented additional initiatives to enhance environmental sustainability, including a low energy air conditioning system and the removal of refrigerators from guest rooms to reduce energy use.

However, these were also identified as areas of concern to guests, and therefore important for management to address to provide consistent service quality. The negative reviews on TripAdvisor revealed that despite the hotel's efforts to mitigate carbon emissions, the customer perceptions of such initiatives were largely negative. This points to the need to advise customers about sustainability initiatives and monitor consumer generated media to mitigate the risks associated with dissatisfied and/or poorly-aware customers, as highlighted in recent research on trust (Filieri et al., 2015).

It is considered a significant lost opportunity that the Sudima Hotel Auckland Airport's management omitted its carbon neutral status on its website, thereby not maximising the benefits of programmes designed to capture 'green' travellers (e.g. TripAdvisor's GreenLeader Program), particularly as their initially stated goal was to attract guests interested in lowering their carbon footprints. Overall, this suggests the need for hotels committed to reducing carbon emissions to carefully communicate their ideology and actions - including in-house policies such as towel re-use - and invest time and resources to respond to consumer generated media such as TripAdvisor reviews. Guests' responses to some initiatives could be easily improved, for example, by explaining that towel reuse is common practice or locating an explanatory card in the bathroom (see Goldstein *et al.*, 2008). This observation resonates with previous research that has underscored the importance of 'green' hotels actively promoting their green campaigns (Han *et al.*, 2010), especially online (Chan, 2013). In this regard, a study by Hsieh (2012) found that less than 50 per cent of hotels utilise websites to communicate their environmental initiatives, suggesting that this may be a widespread problem.

Although some studies (e.g., Pereira-Moliner *et al.*, 2015) noted a positive and significant relationship between environmental management and competitive advantage, businesses are often cautious about launching environmentally focused advertising to promote their green products, in case they are accused of green washing (Peattie and Crane, 2005). To combat these challenges and gain online business, it is important to provide information about sustainable services and products to customers (Zafiropoulos et al., 2006), including details of sustainability initiatives. Hotel guests in this study were poorly aware of such initiatives, although it is acknowledged that this may be because disappointed guests may be more likely than others to leave a public review (Radojevic et al., 2015). Also worthy of comment is the lack of information about the hotel's management plan (Figure 1) and the extent to which it has been achieved. In summary, gaining a competitive advantage from the 'green' positioning of the case hotel was under-utilised, and further investigation into advantages of implementation would benefit both management and guests.

With respect to quality, although the Sudima Hotel Auckland Airport was not marketed as a luxury product, it nevertheless held a 4.5-star Qualmark rating. Guests were provided with all the products and services upon which perceptions of quality are determined (see Kirk, 1995; McCleary et al., 1994; Saleh and Ryan, 1991) – well-appointed clean rooms, comfortable beds, quiet rooms, safety and security, good location, high quality bathroom linen, unlimited water, high pressure showers, amenities such as swimming pools, gyms and saunas, adequate lighting for working and reading, and well-maintained furnishings. Gössling (2011) stated that overall, the more luxurious the accommodation, the more energy will be used, and while this may still be true for many properties, the Sudima Hotel Auckland Airport provides an example of a Qualmark 4.5-star property that does not have high energy use.

Theoretical and Practical Implications

In terms of the nexus between corporate motivations and customer perceptions, this study found that ecological responsibility and competitiveness were the key motivating factors, but that the competitiveness potential was inadequately exploited. This study also reveals that the green motives were not explicitly acknowledged by management, indicating that although it is important for a hotel to undertake mitigation initiatives - specifically to lower emissions – economic imperatives are still deemed important and may not be fully transparent or consciously recognised. This somewhat resonates with previous research, which showed that cost savings were the primary driving factor in implementing environmental initiatives in hotel facilities in Canada (Graci, 2002, Gaci & Dodds, 2008). This 'blurry' area of corporate motivation is visually depicted in Figure 2 by the explicit-implicit continuum.

Figure 2 offers a preliminary framework for informing the motivation-perception juncture in carbon mitigation in the accommodation sector. It is proposed that customer perception and corporate motivation is mediated by communication. In other words, clear communication across a range of channels is likely to determine customer perceptions of corporate motivations with regard to carbon-reduction initiatives. Furthermore, an organisation's core values, actions and philosophy, can be either explicit or implicit, and customers' alignment with these can range from low to high. These conditions may alter the probability of the suggested outcomes. The tentative outcomes articulated in Figure 2 are as follows: (1) clear communication is likely to result in stronger business positioning and competitive advantage, and (2) clear communication is likely to result in more receptive customers, including greater customer satisfaction, retention and experience. Future research can further examine some of the noted nuances, such as the difference between high or low alignment with core values, actions and philosophy on the proposed outcomes (1 and 2). For instance, it would be interesting to know if the satisfaction of customers who support carbon mitigation initiatives increases with better communication, and how this compares with customers who have little interest in carbon reduction. Thus, the information in the Figure 2 can be re-formulated as hypotheses and studied also using quantitative methods, but also examined qualitatively by concentrating on accommodation providers invested in lowering carbon emissions.

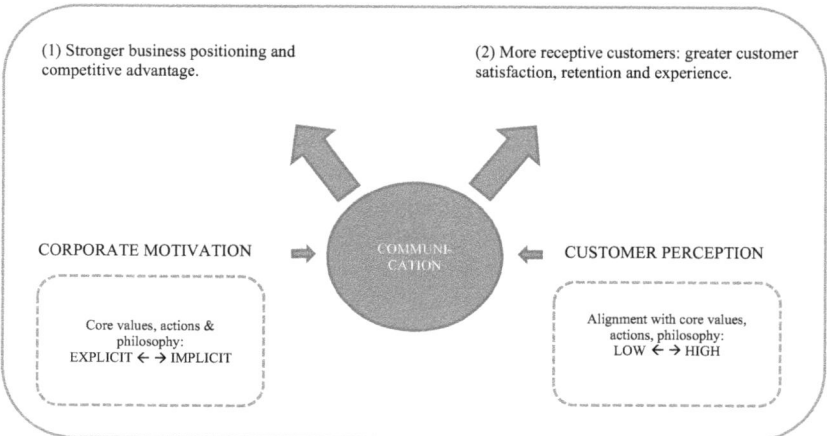

Figure 2: A preliminary framework informing the motivation-perception juncture in carbon mitigation

Overall, Figure 2 provides a step towards a better understanding of the interplay between corporate motivation and customer perception of carbon reduction, making a contribution to theory development in this domain of research. It is imperative to note that this is only an exploratory case study and additional research is needed to test and determine wider applicability. Importantly, the findings in this study are also of practical value to businesses interested in implementing carbon-mitigating initiatives without compromising on comfort and customer satisfaction.

Limitations

This study has several limitations. First, it was necessarily limited to one hotel as there were no other carbon neutral properties in New Zealand during the research period. However, the value of the findings in theory-building case study research lies in teasing out information and relationships that can be used to develop new theoretical frameworks. Future research, including additional case study research, can further assist in sharpening the extent to which the findings may be applied in other contexts.

Another limitation includes the number of interviewees, which were limited to the general manager. Although other managers could have been interviewed, the focus was less on richness, as is common in intrinsic case study research, and more on outlining tentative themes and relationships, and drawing general principles – in alignment with instrumental case studies (Stake, 2005). Finally, due to the limited space and focus, other interconnected areas could not be examined here but can be explored in future research, such as comparisons with studies that have explored the mitigation of other sustainability problems.

References

Bansal, P. and Roth, K. (2000), "Why companies go green: a model of ecological responsiveness", *The Academy of Management Journal*, Vol. 43 No. 4, pp. 717–736.

Barber, N.A. (2014), "Profiling the potential 'green' hotel guest: who are they and what do they want?", *Journal of Hospitality & Tourism Research*, Vol. 38 No. 3, pp. 361–387.

Barber, N.A. and Deale, C. (2014), "Tapping mindfulness to shape hotel guests' sustainable behavior", *Cornell Hospitality Quarterly*, Vol. 55 No. 1, pp. 100–114.

Becken, S. (2013), "Operators' perceptions of energy use and actual saving opportunities for tourism accommodation", *Asia Pacific Journal of Tourism Research*, Vol. 18 No. 1–2, pp. 72–91.

Becken, S., Frampton, C. and Simmons, D. (2001), "Energy consumption patterns in the accommodation sector—the New Zealand case", *Ecological Economics*, Vol. 39 No. 3, pp. 371–386.

Bohdanowicz, P. and Zientara, P. (2008), "Corporate social responsibility in hospitality: issues and implications: a case study of Scandic", *Scandinavian Journal of Hospitality and Tourism*, Vol. 8 No. 4, pp. 271–293.

Brundtland, G.H. (1987), "Our Common Future", presented at the United Nations, The World Commission on Environment and Development, available at: http://conspect.nl/pdf/Our_Common_Future-Brundtland_Report_1987.pdf. (accessed 29 March 2017).

Bruns-Smith, A., Choy, V., Chong, H. and Verma, R. (2015), "Environmental sustainability in the hospitality industry: best practices, guest participation, and customer satisfaction.", *Cornell Hospitality Report*, Vol. 15 No. 3, pp. 6–16.

Budeanu, A., Miller, G., Moscardo, G. and Ooi, C.-S. (2016), "Sustainable tourism, progress, challenges and opportunities: an introduction", *Journal of Cleaner Production*, Vol. 111, pp. 285–294.

Chan, E.S.W. (2013), "Gap analysis of green hotel marketing", *International Journal of Contemporary Hospitality Management*, Vol. 25 No. 7, pp. 1017–1048.

Chen, M.-F. & Tung, P.-J., 2014. Developing an extended theory of planned behavior model to predict consumers' intention to visit green hotels. *International Journal of Hospitality Management*, 36, 221–230.

Chou, C.-J. & Chen, P.-C., 2014. Preferences and willingness to pay for green hotel attributes in tourist choice behavior: the case of Taiwan. *Journal of Travel & Tourism Marketing*, 31, 937–957.

Creswell, J.W. and Plano Clark, V.L. (2007), *Designing and Conducting Mixed Methods Research*, Sage Publications, Thousand Oaks, CA.

Davies, B. (2003), "The role of quantitative and qualitative research in industrial studies of tourism", *International Journal of Tourism Research*, Vol. 5 No. 2, pp. 97–111.

de Grosbois, D. and Fennell, D. (2011), "Carbon footprint of the global hotel companies: comparison of methodologies and results", *Tourism Recreation Research*, Vol. 36 No. 3, pp. 231–245.

Denzin, N. K., & Lincoln, Y. S. (2011). *The SAGE handbook of Qualitative Research*

(4th ed.). Thousand Oaks, CA: SAGE Publications, Inc.

el Dief, M. and Font, X. (2010), "The determinants of hotels' marketing managers' green marketing behaviour", *Journal of Sustainable Tourism*, Vol. 18 No. 2, pp. 157-174.

Ecolab (2016), available at: http://en-nz.ecolab.com/ (accessed 29 March 2017).

Eisenhardt, K. M. (1989). Building theories from case study research. *The Academy of Management Review*, Vol. 14 No. 4, pp. 532-550.

Feilzer, M. Y. (2010). Doing mixed methods research pragmatically: implications for the rediscovery of pragmatism as a research paradigm. *Journal of Mixed Methods Research*, Vol. 4 No. 1, pp.6-16.

Filieri, R., Alguezaui, S. and McLeay, F. (2015), "Why do travellers trust TripAdvisor? antecedents of trust towards consumer-generated media and its influence on recommendation adoption and word of mouth", *Tourism Management*, Vol. 51, pp. 174-185.

Gillham, B. (2000), *Case Study Research Methods*, Continuum, London, UK.

Goldstein, N.J., Cialdini, R.B. and Griskevicius, V. (2008), "A room with a viewpoint: using social norms to motivate environmental conservation in hotels", *Journal of Consumer Research*, Vol. 35 No. 3, pp. 472-482.

Gössling, S. (2011), *Carbon Management in Tourism: Mitigating the Impacts on Climate Change*, Routledge, Oxon, UK.

Graci, S. (2002). The greening of accommodation: A study of voluntary environmental initiatives in the hotel industry. Toronto: University of Toronto.

Graci, S., & Dodds, R. (2008). Why go green? The business case for environmental commitment in the Canadian hotel industry. *Anatolia*, 19(2), 251-270. doi: 10.1080/13032917.2008.9687072

Hall, Amelung, B., Cohen, S., Eijgelaar, E., Gössling, S., Higham, J., Leemans, R., et al. (2015), "On climate change skepticism and denial in tourism", *Journal of Sustainable Tourism*, Vol. 23 No. 1, pp. 4-25.

Han, X. & Chan, K., 2013. Perception of green hotels among tourists in Hong Kong: An exploratory study. *Services Marketing Quarterly*, 34, 339-352.

Han, H., Hsu, L.-T. (Jane) and Sheu, C. (2010), "Application of the theory of planned behavior to green hotel choice: testing the effect of environmental (sic) friendly activities", *Tourism Management*, Vol. 31 No. 3, pp. 325-334.

Hsieh, Y. C. (2012). Hotel companies' environmental policies and practices: a content analysis of their web pages. *International Journal of Contemporary Hospitality Management*, 24(1), 97-121. doi: 10.1108/095961112

Heardable (2012), "The world's 10 best hotel chains", *Online Brand Health Benchmark Report*, available at: https://www.samselocity.com/uploads/2/5/6/9/25696746/the-worlds-10-best-hotel-chains.pdf (accessed 27 April 2018).

Hsieh, H.-F., and Shannon, S. (2005). "Three approaches to qualitative content analysis", *Qualitative Health Research*, Vol. 15 No. 9, pp. 1277–1288.

Johnson, R. B., Onwuegbuzie, A. J., & Turner, L. A. (2007). "Toward a definition of mixed methods research", *Journal of Mixed Methods Research*, Vol. 1 No. 2, pp. 112–133.

Jones, P., Hillier, D. and Comfort, D. (2014), "Sustainability in the global hotel industry", *International Journal of Contemporary Hospitality Management*, Vol. 26 No. 1, pp. 5–17.

Kasim, A., 2004. Socio-environmentally responsible hotel business: Do tourists to Penang Island, Malaysia care? *Journal of Hospitality & Leisure Marketing*, 11, 5–28.

Kim, J., Braun, J., Tzempelikos, A. and Horton, W.T. (2016), "Performance evaluation of a passive chilled beam system and comparison with a conventional air system", presented at the International High Performance Buildings Conference, Purdue e-pubs, available at: http://docs.lib.purdue.edu/ihpbc/179 (Accessed 29 March 2017).

Kirk, D. (1995), "Environmental management in hotels", *International Journal of Contemporary Hospitality Management*, Vol. 7 No. 6, pp. 3–8.

Kuminoff, N. V., Zhang, C. & Rudi, J., 2010. Are travelers willing to pay a premium to stay at a green hotel? Evidence from an internal meta-analysis of hedonic price premia. *Agricultural and Resource Economics Review* 39, 468–484.

Legrand, W. and Nielsen, R.S. (2017), "Climate-conscious identity and climate-adaptive innovations in hospitality", *Advances in Hospitality and Leisure*, Vol. 13, pp. 63–78.

Leslie, B. (2015), "Auckland hotel achieves carboNZero first", *Accomnews NZ*, available at: http://www.accomnews.co.nz/2015/07/08/auckland-hotel-achieves-carbonzero-first/ (accessed 9 October 2016).

Levinas, E., 1991. *Otherwise than being or beyond essence*. Dordrecht, The Netherlands, Kluwer Academic Publishers.

Line, N.D. and Hanks, L. (2016), "The effects of environmental and luxury beliefs on intention to patronize green hotels: the moderating effect of destination image", *Journal of Sustainable Tourism*, Vol. 24 No. 6, pp. 904–925.

Manaktola, K. and Jauhari, V. (2007), "Exploring consumer attitude and behaviour towards green practices in the lodging industry in India", *International Journal of Contemporary Hospitality Management*, Vol. 19 No. 5, pp. 364–377.

Maxwell, J.A. (2010), "Using numbers in qualitative research", *Qualitative Inquiry*, Vol. 16 No. 6, pp. 475–482.

McCleary, K.W., Weaver, P.A. and Lan, L. (1994), "Gender-based differences in business travelers' lodging preferences", *The Cornell Hotel and Restaurant Administration Quarterly*, Vol. 35 No. 2, pp. 51–58.

Michailidou, A.V., Vlachokostas, C. and Moussiopoulos, N. (2016), "Interactions between climate change and the tourism sector: multiple-criteria decision analysis to assess mitigation and adaptation options in tourism areas", *Tourism Management*, Vol. 55, pp. 1–12.

Mills, A. J., Durepos, G., & Wiebe, E. (Eds.). (2010). *Encyclopedia of case study research*. Thousand Oaks, CA: SAGE Publications.

Ministry of Business, Innovation and Employment (2018), available at: http://www.mbie.govt.nz/info-services/sectors-industries/tourism/tourism-research-data/tourism-satellite-account (accessed 1 April 2018).

Nadkarni, D. (2011), "Subtlety, elegance imbue Sudima's new Auckland hotel", *Indian Weekender*, New Zealand.

Organisation for Economic Cooperation and Development (2017), "Environmental pressures rising in New Zealand", available at: http://www.oecd.org/newzealand/ (accessed 13 March 2018).

Okereke, C. (2007), "An exploration of motivations, drivers and barriers to carbon management: the UK FTSE 100", *European Management Journal*, Vol. 25 No. 6, pp. 475–486.

Peattie, K. and Crane, A. (2005), "Green marketing: legend, myth, farce or prophesy?", *Qualitative Market Research: An International Journal*, Vol. 8 No. 4, pp. 357–370.

Pereira-Moliner, J., Font, X., Tarí, J.J., Molina-Azorin, J.F., Lopez-Gamero, M.D. and Pertusa-Ortega, E.M. (2015), "The holy grail: environmental management, competitive advantage and business performance in the Spanish hotel industry", *International Journal of Contemporary Hospitality Management*, Vol. 27 No. 5, pp. 714–738.

Plaza (2014), "Three-year-old Auckland hotel achieves carbonzero status in a New Zealand first", *Tourism Export Council of New Zealand*, available at: http://www.tourismexportcouncil.org.nz/three-year-old-auckland-hotel-achieves-carbonzero-status-in-a-new-zealand-first/ (accessed 23 June 2016).

Prairie, P. (2012), "A critical look at hotel sustainability", *The Huffington Post*, available at: http://www.huffingtonpost.com/patti-prairie/a-critical-look-at-hotel-_b_1471188.html (accessed 10 September 2016).

Qualmark (2018), "How Qualmark works", available at: https://www.qualmark.co.nz/en/learn-about-us/how-qualmark-works (accessed 27 April 2018).

Radojevic, T., Stanisic, N. and Stanic, N. (2015), "Ensuring positive feedback: factors that influence customer satisfaction in the contemporary hospitality industry", *Tourism Management*, Vol. 51, pp. 13–21.

Rauch, D., Collins, D., Nale, R. and Barr, P. (2015), "Measuring service quality in mid-scale hotels", *International Journal of Contemporary Hospitality Management*, Vol. Vol 27 No. 1, pp. 87–106.

Ricaurte, E. (2011), "Developing a sustainability measurement framework for hotels: toward an industry-wide reporting structure", *Center for Hospitality Research Publications*, Vol. 11 No. 13, pp. 6–30.

Roberts, J., 2003. The manufacture of corporate social responsibility: Constructing corporate sensibility. *Organization*, 10, 249–265.

Rust, R.T. and Oliver, R.L. (1994), "Service quality: insights and managerial implications from the frontier", *Service Quality: New Directions in Theory and Practice*, SAGE Publications, Thousand Oaks, CA, pp. 1–20.

Saleh, F. and Ryan, C. (1991), "Analysing service quality in the hospitality industry using the servqual model", *The Service Industries Journal*, Vol. 11 No. 3, pp. 324–345.

Scott, D., Gössling, S. and Hall, C.M. (2012), "International tourism and climate change", *Wiley Interdisciplinary Reviews: Climate Change*, Vol. 3 No. 3, pp. 213–232.

Singh, N., Cranage, D. and Lee, S. (2014), "Green strategies for hotels: estimation of recycling benefits", *International Journal of Hospitality Management*, Vol. 43, pp. 13–22.

Su, Y.-P., Hall, C.M. and Ozanne, L. (2013), "Hospitality industry responses to climate change: a benchmark study of Taiwanese tourist hotels", *Asia Pacific Journal of Tourism Research*, Vol. 18 No. 1–2, pp. 92–107.

Sudima Auckland Airport Hotel (2018), *TripAdvisor*, available at: http://www.tripadvisor.co.nz/Hotel_Review-g255580-d2038909-Reviews-Sudima_Auckland_Airport_Hotel-Mangere_North_Island.html (accessed 1 April 2018).

Sudima Hotels (2018), available at: /about/environmental-sustainability/ (accessed 23 April 2018).

Stake, R. E. (2005). *Qualitative Case Studies*. In N. K. Denzin & Y. S. Lincoln (Eds.), The SAGE handbook of qualitative research (3rd ed., pp. 443–466). Thousand Oaks, CA: SAGE.

Teddlie, C. and Tashakkori, A. (2009), *Foundations of Mixed Methods Research Integrating Quantitative and Qualitative Approaches in the Social and Behavioural Sciences*, Thousand Oaks, CA: SAGE.

Tourism New Zealand (2018a), "About the industry", available at: https://www.tourismnewzealand.com/about/about-the-industry/ (accessed 27 April 2018).

Tourism New Zealand (2018b), "What we do", available at: https://www.tourismnewzealand.com/about/what-we-do/campaign-and-activity/ (accessed 27 April 2018).

The World Bank (2018), "World Bank countries and lending groups", available at: https://datahelpdesk.worldbank.org/knowledgebase/articles/906519 (accessed 27 April 2018).

UNWTO (2015), "UNWTO Tourism Highlights 2015 Edition", available at: http://www.e-unwto.org/doi/pdf/10.18111/9789284416899 (accessed 28 March 2017).

UNWTO (2017), available at: http://sdt.unwto.org/content/faq-climate-change-and-tourism (accessed 28 March 2017).

UNWTO (2018), available at: http://media.unwto.org/press-release/2018-01-15/2017-international-tourism-results-highest-seven-years (accessed 1 April 2018).

Vernon, J., Essex, S., Pinder, D. and Curry, K. (2003), "The 'greening' of tourism micro-businesses: outcomes of focus group investigations in South East Cornwall", *Business Strategy and the Environment*, Vol. 12 No. 1, pp. 49–69.

Watertech Plus (2016), available at: http://www.watertechplus.co.nz/ (accessed 29 March 2017).

World Tourism Organisation and United Nations Environment Programme (2008), *Climate Change and Tourism - Responding to Global Challenges*, UNWTO and UNEP, Madrid, Spain, available at: http://sdt.unwto.org/sites/all/files/docpdf/climate2008.pdf (Accessed 27 March 2017).

Yang, W. (2010), "The development of tourism in the low carbon economy", *International Business Research*, Vol. 3 No. 4, pp. 212–215.

Zafiropoulos, C., Vrana, V. and Paschaloudis, D. (2006), "The internet practices of hotel companies: an analysis from Greece", *International Journal of Contemporary Hospitality Management*, Vol. 18 No. 2, pp. 156–163.

Zero Carbon Act (2018), available at: https://www.mfe.govt.nz/news-events/zero-carbon-act (accessed 21 March, 2018).

Sabrina Seeler, Michael Lück and Heike Schänzel

Increased travel experience and its effects on responsible and sustainable travel behaviour

1. Introduction

With the sustained growth in international tourist arrivals worldwide (World Tourism Organization (UNWTO), 2018a), markets are maturing and tourists are proposed to be increasingly experienced. As suggested by previous research, higher levels of past travel experience contribute to changes in travel decision-making, travel behaviour and motivation (Lehto, O'Leary, & Morrison, 2004; Pearce & Caltabiano, 1983; Pearce & Lee, 2005). With their higher levels of past travel experiences, tourists are increasingly looking for self-fulfilment and individualisation, are desiring authentic and immersive experiences and are thought to travel more independently, sustainably and responsibly. Alongside these general changes, tourist demand has become more diversified and tourists exhibit an increasingly hybridised travel behaviour that challenges tourism marketers to segment tourists into internally-homogenous and externally-heterogeneous groups (Boztug, Babakhani, Dolnicar, & Laesser, 2015). Csikszentmihalyi and Coffey (2017) further noted that the travel behaviour and motivation of tourists not only varies between trips but is also often fluid and evolving during a trip.

With the aim of catering to the demands of increasingly experienced and hybrid tourists, the four realms of experience creation were introduced in the tourism context (Pine & Gilmore, 2011). In regard to the creation of experiences in tourism, it remains unanswered how an individual's level of experience is reflected in his or her travel behaviour and how this, in turn, impacts the creation and delivery of experiences that encourage sustainable travel development. The tourist destination plays a pivotal role in the creation and delivery of these experiences and represents the *experiencescape* for the tourist (Mossberg, 2007). With the expanding number of tourist destinations worldwide, competition grows and aspects of sustainability gain centre stage in strategic destination management (López-Sánchez & Pulido-Fernández, 2016). Competitive advantages can only be sustained when the interconnectedness between the supply-side and demand-side is fully comprehended, and a macro-micro perspective is established to understand the demand of today's increasingly experienced tourists. As part of a broader study, this research project aimed at closing existing gaps in the academic literature by addressing the following two research objectives:

1. To understand the experienced tourists' travel behaviour from a macro-micro perspective.
2. To determine whether there are differences between tourists with different experience levels regarding sustainable travel behaviour.

Based on results from an exploratory, sequential mixed-methods study conducted in Germany and New Zealand, this chapter provides an overview of the findings related to these research questions.

2. The concept of experience in the tourism context

Past literature on the concept of experience in the tourism context demonstrates that two broad subtopics are commonly examined: (1) experience creation in tourism; and (2) experience consumption in tourism (Seeler, Lück, & Schänzel, 2018). A brief overview of these two subtopics is presented in the following sections.

2.1 *Experience creation in tourism*

Pine and Gilmore (1999) introduced the 'four realms of experience' by distinguishing between active and passive participation as well as absorption and immersion, leading to the proposal of the four realms: entertainment, educational, esthetical and escapist. By incorporating the degree of customer involvement and degree of relationship with the external environment in their model, Pine and Gilmore (1999) emphasised the mutual importance of both the supplier and the consumer of the experience. The tourism product was initially not the focus of the model, yet several scholars adopted the concept of the experience economy in the context of tourism (Chang, 2018; Oh, Fiore, & Jeoung, 2007). Despite crediting the consumer with central importance in the realisation of the tourism experience, most studies have adopted a marketing/managerial perspective and explored the underlying drivers to delivering positive memorable experiences. Social science approaches that consider positive memorable experiences more holistically and beyond the actual on-site experience created by the tourism industry itself are lacking (Quan & Wang, 2004; Tung & Ritchie, 2011).

Social scientists have emphasised that the metaphysical dimensions of space and time need to be considered before a comprehensive understanding of the creation and consumption of positive memorable experiences can be gained

(Ek, Larsen, Hornskov, & Mansfeldt, 2008; Tung & Ritchie, 2011). In terms of the temporal element, scholars proposed that pre-trip expectations, on-trip experiences and post-trip recollections need to be equally acknowledged (Kruger & Saayman, 2017). Alongside the tourist destination as the space where the on-trip experiences are consumed, general/non-tourist experiences are also consumed in the usual daily-life environment and contribute to the creation of the actual trip experience. However, scholars generally agreed that the experiences lived through at the destination are the most influential (Ek et al., 2008). Thus, the tourist destination gains centre stage in the creation and delivery of tourism experiences and the tourist destination is upgraded to an *experiencescape* (Mossberg, 2007). Despite the central role of the tourist destination and tourism suppliers within the complex destination network in providing the stage for the experience, the physical body and mind of a tourist is required to translate these inputs into actual lived experiences. Consequently, more emphasis needs to be given to tourists themselves and the process of experience consumption and, eventually, accumulation of travel experiences.

2.2 *Experience consumption in tourism*

From a more subjective social science approach, tourists are no longer limited to being passive receivers and spectators; instead, they are identified as active co-creators, co-producers and 'experiencers' (Bosangit, Hibbert, & McCabe, 2015; Mossberg, 2007). The process of experiencing becomes a subjective inner process and a deeply rooted psychological phenomenon that can only be encoded by the individuals themselves (Chen, Prebensen, & Uysal, 2014). These deeply rooted subjective processes reinforce the accumulation of experiences and encourage the development and transformation of self (Bosangit et al., 2015; Jantzen, 2013). Tourist destinations provide physical spaces and services that are consumed by tourists to stimulate these inner processes. Despite the acknowledgement of the mutual dependencies and dynamic relationships between the supply-side and demand-side of tourism, it remains unanswered which drivers eventually contribute to the meaning-making processes of experience accumulation.

The accumulation of experience not only contributes to personal change and the transformation of self-identity, the changes in the level of prior experiences also impact an individual's motivation to travel and their travel behaviour. Scholars have addressed the relationships between prior travel experiences and motivation to travel, travel decision-making, travel behaviour and destination choice and suggested models to predict particular aspects of the travel journey based on the level of an individual's prior experience (Lehto et al., 2004; Pearce & Caltabiano, 1983). Most of these scholars referred to a few quantifiable factors in order to predict the level of prior experience, such as the age of the respondent or the number of domestic and/or international trips. One of the most cited and widely applied models is Pearce and colleagues' travel career ladder (TCL), later refined as travel career pattern (TCP; Pearce & Caltabiano, 1983; Pearce & Lee, 2005).

There is consensus that prior experience and knowledge impact travel motivation and behaviour, yet while prior knowledge has been considered a multidimensional construct (Sharifpour & Walters, 2014), most studies that address the level of prior experience to predict travel behaviour and motivation refer to unidimensional measurement tools. As the level of prior experience is considered a precursor to motivation and behaviour, simplified measurement tools are applied that mostly rely on predefined, countable variables. What remains unresolved in contemporary academic literature is whether these countable factors can reliably classify tourists in different experience level categories.

A more comprehensive understanding of the drivers that eventually impact the process of experience accumulation is needed. Instead of treating the level of experience as an antecedent, a step back is required to holistically understand which factors contribute to higher experience levels, with the level of experience being treated as an outcome variable instead. Without an unambiguous understanding of the factors that influence the process of experience accumulation, which then identifies an experienced tourist, the predictability of any aspect of the travel journey based on the individual's level of experience is lacking evidence. Once the distinguishing features of an experienced tourist are known, further analysis is feasible to investigate whether there are differences in travel behaviour by subgroups.

3. Changes in travel behaviour towards sustainability

Environmental awareness, ethical consciousness and responsibility have experienced sustained growth in the domain of consumer decision-making and behaviour in the recent past (Fennell, 2006; Holden, 2016). Consequently, research in this area has expanded and different approaches and perspectives have been adopted to examine the sustainable behaviour of consumers (Lanzini, 2018; van Trijp, 2014). With the growing number of international tourists worldwide, the tourism industry has become a major driver of global economic and social well-being (World Travel & Tourism Council, 2017). At the same time, the growth in international tourism has been accompanied by an increase of CO_2 emissions from tourism, which heavily impacts climate change (Hall, 2013). More recently, the growth in international tourism has resulted in overcrowding and increasing pressure on some major tourist destinations. This new phenomenon has been identified as *overtourism* (Muler Gonzalez, Coromina, & Galí, 2018; Postma & Schmuecker, 2017). Scholars have discussed the centrality of residents in developing sustainable tourism destinations and raised the concern of social carrying capacity and changes in residents' perceptions of tourism, particularly in towns and cities with high visitor pressure. Issues around overtourism amplify the need for sustainable strategies that take tourist dispersion into consideration, so that tourism demand is more spread out both within and across destinations.

The tourism industry is obligated to continuously work towards more sustainable tourism practices and to align tourism policies and business operations with sustainable development goals, as defined by the World Tourism Organization and the United Nations Development Programme (UNWTO & UNDP, 2017). In early 2018, the UNWTO rebranded the 10-Year Framework of Programmes on Sustainable Consumption and Production (10YFP) as a One Planet network with the aim of advancing sustainable consumption and production in the tourism industry (UNWTO, 2018b). Thereby, the UNWTO clearly emphasised the contribution of both tourism suppliers (production) and tourists (consumers).

Higher levels of prior experiences have been associated with greater knowledge and awareness regarding the economic, environmental and social impact of tourism. It is hoped that this increased awareness will be translated into more responsible and sustainable travel behaviour (Lee, Bonn, Reid, & Kim, 2017). An unambiguous definition of responsible, sustainable travel behaviour is missing to date, but the United Nations Environment Programme (UNEP) and UNWTO have defined sustainable tourism as follows:

> Tourism that takes full account of its current and future economic, social and environmental impacts, addressing the needs of visitors, the industry, the environment and host communities. (UNEP & UNWTO, 2005, p. 12)

It is further stated that sustainable tourism is not a special form of tourism; instead, sustainable practices need to be incorporated in all forms of tourism development (Lück, 2002). Most existing studies have focussed on environmental concerns (especially climate change), addressed particular aspects of behaviour (e.g., mobility) and focussed on one particular area of daily life (e.g., the home environment, daily commute, or holiday travel) (Thøgersen & Ölander, 2003). While research on sustainable behaviour has been limited to rather discrete aspects (e.g., sustainable mobility behaviour) and failed to comprehensively define what sustainable travel behaviour entails, the overall aim of most studies has been to examine behavioural change towards more sustainable practices.

Several scholars have critically scrutinised the complexity of behavioural changes regarding holiday travel behaviour. While some scholars have argued that behavioural changes in one area of life increase the likelihood of changes in other domains, other researchers have postulated that correlations between different areas of life are limited and can be negative (Margetts & Kashima, 2017; Thøgersen & Ölander, 2003). This adverse spill-over effect was summarised by Prillwitz and Barr (2011) as: "the performance of an environmentally friendly behaviour in one area can also reduce the propensity to behave environmentally friendly in other areas: 'Good' in one domain of individual life is sometimes used to justify 'bad' behaviour in another domain" (p. 1592). It was claimed that while travelling on holiday, the tolerance of 'bad' behaviour is higher, and the implementation of behavioural changes are at a reduced level compared to changes in the context of the home environment (e.g., food waste, responsible eating). Various factors have been theorised to account for this reluctance to make behavioural change while travelling on holiday, and two major areas of concern have emerged in the literature to explain the lack of transforming pro-sustainable attitudes into actual travel behaviour: (1) the attitude–behaviour gap; and (2) the lack of information and sustainable offers (Gössling & Peeters, 2007; Juvan & Dolnicar, 2014).

Several authors have found that the general increased awareness towards more sustainable consumption decisions translates only to a limited degree to actual purchase and consumption behaviour (Antimova, Nawijn, & Peeters, 2012; Juvan & Dolnicar, 2014; Lee et al., 2017). Van Trijp (2014) summarised this discrepancy as follows: "Although they 'talk green', they do not necessarily 'walk green'" (p. 4). Among other theories, scholars have frequently referred to the theory of planned behaviour (Ajzen, 1985) to encourage a better understanding of behaviour and non-behaviour. However, it was found that tourists tend to have a lower perception of responsibility when travelling, as they understand their holiday travel as a special time of the year in which they want to feel free from responsibilities and are less willing to change behaviour. In their comprehensive research on reasons for the attitude–behaviour gap in sustainable travel behaviour, Juvan and Dolnicar (2014) summarised this as "vacations are an exception" (p. 90). These authors also found a "denial of control" (p. 89) among respondents who stated that they would travel more sustainably if they had the financial resources, enough information and the necessary infrastructure.

Scholars confirmed that the external environment and supply-side of tourism are lacking alternative options and do not comprehensively provide information to boost public awareness (Antimova et al., 2012; López-Sánchez & Pulido-Fernández, 2016). Thus, economic tools and the means for more appropriate infrastructure are needed to encourage more sustainable travel behaviour and habits. In the first place, it is about the creation of awareness and the provision of information. This requires voluntary and informed participation and the long-term commitment of all key players, including governments. However, research has also shown that being confronted with massive global environmental concerns has an adverse effect on consumers, as messages can be intimidating and induce people to maintain the *status quo* instead of changing their habits and behaviour (Banfield, Shepherd, & Kay, 2014). Instead of limiting travel research to very specific and obvious forms of sustainable travel, such as environmentally friendly transport options, more studies are required that take less specific forms of sustainable travel into consideration. These softer forms of sustainability equally contribute to the mitigation of environmental, social or economic concerns; examples are travelling off the beaten path and the readiness to learn about new cultures and regions, which both encourage dispersion and alleviate pressure in places of high tourism demand. It is these softer forms of sustainable travel behaviour and their relationship with experienced tourists that were examined in this study.

4. Research methodology and methods

Due to the complexity of the phenomenon being studied, a single method approach in one country would have been insufficient. Therefore, an exploratory, sequential mixed-methods research design was employed in two case studies, Germany and New Zealand. Prior to the implementation of the study, the level of interaction between the two research phases, the relative priority, timing and procedures needed to be identified (Creswell & Plano Clark, 2011). As is common in exploratory mixed-methods approaches, a qualitative phase was introduced first followed by a quantitative research phase, which was dependent on the completion of the first phase. Thus, the timing was sequential and the purposes of mixing methods were complementary and developmental (Denscombe, 2017).

The exploratory, sequential mixed-methods approach comprised two data collection instruments: semi-structured interviews and a web panel online survey. The semi-structured interviews were conducted with 18 industry experts from 15 Destination Marketing Organisations (DMOs) across Germany and New Zealand. Participants were selected using a purposive sampling strategy. The interviews were carried out face-to-face in the respective organisational premises of the DMOs between May and August 2016. All interviews were audio-recorded and transcribed after completion, and the transcripts were confirmed by the research participants. Data were analysed following Braun and Clarke's (2006) thematic analysis and a two-cycle coding process (manual and digital coding) using NVivo 11 was adopted. Once the qualitative data analysis was completed, the questionnaire for the online survey was developed and the second research phase initiated.

The target population for the second, quantitative research phase was defined as Germans and New Zealanders aged between 20 and 69 years with any past travel experiences. A nonprobability quota sampling strategy was applied, and respondents were accessed through a web panel provider. Based on official census data, the demographic factors of age, gender and region were used to define the interlocked quota for data collection and the total sample size per case study was 500 respondents. Data collection took place in January 2017 and the self-administered online survey was completed after seven days of fieldwork. After data processing, data cleaning, and tests for reliability and validity, univariate (descriptive statistics) and bivariate (cross-tabs, Pearson's Chi-square) were carried out using IBM SPSS, version 23.

The exploratory, sequential mixed-methods research design enabled the incorporation of the supply-side (DMO representatives) and the demand-side of tourism (tourists) and established a macro-micro perspective. The multi-stakeholder perspective was further expanded by the implementation of both research phases in the two case studies and facilitation of a cross-national comparative perspective.

5. Findings

5.1 *Qualitative findings – the supply-side perspective*

The findings from the semi-structured interviews with representatives from 15 DMOs across Germany and New Zealand are presented in the following sections and supported by direct quotes from the individual interviews. For reasons of participant confidentiality, pseudonyms were assigned to each participant and these have been used in the sections that follow.

Distinguishing experienced from less experienced tourists

Industry representatives from German and New Zealand DMOs agreed that today's tourists are increasingly experienced and proposed several factors that contribute to the process of experience accumulation. Findings revealed eight independent, yet partly interrelated, dimensions that were grouped in two themes to identify an experienced tourist: personal identifiers and external facilitators. Several highly subjective dimensions referred to an individual's profiling information and were labelled as 'personal identifiers'; these encompassed sociodemographic information, social identity elements, life stage and travel biography. Although the quantity of past travel was highlighted as a potential driver behind higher experience levels of tourists, destination representatives strongly agreed that being identified as an experienced tourist goes beyond purely quantifiable factors. Instead, it was emphasised that experienced tourists will have been exposed to different cultures and customs, will have had more awareness of the places they travelled to and been driven by motives that went beyond hedonic reasons and instead travelled for self-fulfilment and growth.

> *Max (GER): Well, quantity – there are people that travel to the same destination and resort for thirty years and will have thirty years of travel experience in that particular destination and there are people that have travelled for thirty years but have visited 60 different travel destinations worldwide.*

Destination marketers further suggested that exhibiting independent travel behaviour and identifying as a free independent traveller (FIT) are central and decisive factors in distinguishing an experienced from a less experienced tourist.

> *Jordan (NZ): I guess, the nature of the travellers is less inclined to be FIT the less experienced they are.*

In addition, industry experts proposed external environment factors that facilitate the process of experience accumulation. Several industry experts discussed the technological advancements and development of online media as external facilitators of the accumulation of travel experience. Technology provides not only the opportunity to gather information more quickly, it also delivers the ready comparison of travel options and prices and increased transparency facilitated through user-generated content and review sites. It was also argued that travel becomes easier as the individual's smart phone is an information source, travel guide and navigation device in one.

> *Matt & George (NZ): I think it is the availability of information, the instant availability and it gives them, TripAdvisor and so on, give them the confidence that they will be ok.*

Alongside the accessibility of information, destination representatives commented on the continuous importance of personal reference groups and word-of-mouth, which is further accelerated through online social network sites and sharing platforms.

> *John (NZ): A lot of those thoughts are seeded by their friends' social media activity and sharing of their own holiday experiences. Word-of-mouth has been the best form of advertising since day one and it hasn't changed. What has changed is how it gets out there.*

Destination experts from Germany and New Zealand shared the understanding that the process of experience accumulation is multidimensional and agreed that an individual's level of experience influences their travel behaviour and positively contributes to more sustainable and responsible travel forms.

Experienced tourists and sustainable travel behaviour

Industry experts proposed that tourists with higher levels of prior experience show a greater interest in the destination itself and want to get in touch with local communities. They are looking for immersion, want to get engaged and seek authentic experiences. It was also suggested that experienced tourists are willing to travel off the beaten path to satisfy their desires for more real and authentic experiences. In this way they may contribute to dispersion from frequently visited, major sites and help to alleviate overcrowding. They are willing to invest more time in understanding a region and its people and are ready to pay a higher price for these experiences. This is shown in the following quotes:

> *Hannah & Carl (GER): I am willing to pay money to get to know a region better. I prefer sitting on an improvised table and have less for breakfast, but I eat what people from the region eat for breakfast.*

> *Harry & Ann (NZ): They actually want to sit down, discuss with you, and take your time to actually understand it.*

Thereby, experts proposed that experienced tourists want to leave positive footprints. Their increased knowledge about the places they visit and consciousness about the impact of tourism encourages them to positively contribute to the well-being of a destination.

> *Lara (NZ): And it might be that they go to a struggling nation and they can actually do some goodwill over there and they feel good.*

In addition to having increased knowledge prior to the travel experience, it was assumed that more experienced tourists aim to continue learning, broaden their horizons and grow as persons.

> Rachel (NZ): The reason people travel is about having those experiences that shape you as an individual and learning about different cultures, and meeting people.

Although there was consensus that tourists with higher levels of past travel experiences have higher degrees of consciousness and are more sensitive in regard to their own responsibility as tourists, it was questioned whether this behaviour reflects pseudo or genuine sustainable travel behaviour. This is exemplified in the following quote:

> Fred (GER): Market research tells us that guests consider issues of sustainability as very important criterion in decision-making in tourism. Looking at the actual acceptance and booking of concrete offers, it is only limitedly recognisable.

The German experts compared this pseudo behaviour with an environmental campaign in Germany in the late 1980s. People were asked whether they would invest 200 or 300 German Marks to contribute to nature conservation and more than 90 percent stated that they would. However, when they were asked to upgrade their car with a catalytic converter, only a small percentage of the population actually did that. Here, the destination representative referred to social desirability in relation to aspects of sustainability and nature conservation and emphasised that despite the generally higher level of past travel experience and the growing number of experienced tourists, only a small group of these tourists change their travel behaviour by accepting and booking offers of more responsible and sustainable travel.

These responsible and sustainable travel forms were more evident in the travel behaviour of younger generations, as the desire for sustainable and responsible travel is associated with their changed value systems. Several experts noted that these changes are particularly noticeable in the shifting food habits towards regional and organic produce. It was further suggested that the younger generation carries a greater responsibility towards their children and values authentic experiences, closeness to nature, and aspects of sustainability. These values were believed to influence the accumulation of travel experiences and eventually the travel behaviour of more experienced tourists. Whether and to what degree the destination experts' perceptions of the dimensions of experienced tourists and their travel behaviour is reflected in the tourists' realities was investigated in the second research phase.

5.2 *Quantitative findings – the demand-side perspective*

Findings from the quantitative online survey are reported in the following sections with results from the German and New Zealand samples introduced

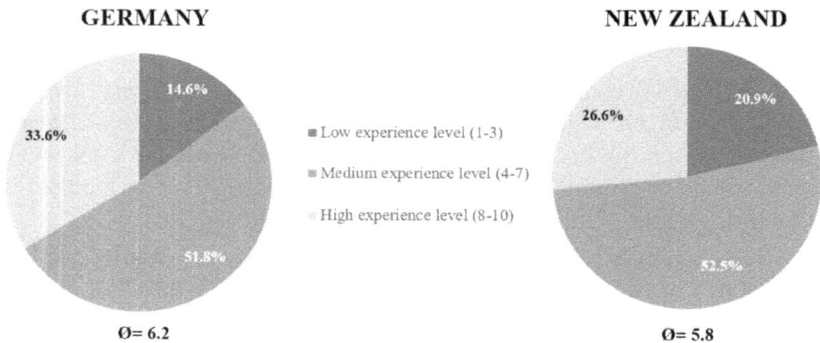

Figure 1: Self-rated travel experience level of tourists in Germany and New Zealand (Ø=average rating)

separately. The aim of this chapter is to investigate whether differences exist between respondents with greater and less past travel experience. Emphasis was given to questions that were related to responsible and sustainable travel behaviour.

Describing the German and New Zealand tourist by self-rated experience level

Respondents were asked to self-rate their level of travel experience on a scale from 1 (*not experienced at all*) to 10 (*extremely experienced*). On average, Germans rated their level of experience as 6.2 whereas the average rate of New Zealanders was slightly lower (5.8). Based on their self-rated experience level respondents were segmented into three groups: low experience (1–3), medium experience (4–7), and high experience (8–10). The distribution per case study is visualised in Figure 1.

Cross tabulations with Pearson's Chi-square were computed to gain an overall picture of the respondents' socio-demographic profiles by country subgroup. A level of confidence of 95 % was applied, corresponding to a margin of error of ±5 % and only statistical significant results ($p \leq .05$) are reported (Alreck & Settle, 2004). In both countries, there was a greater percentage of male respondents with a self-rated high level of travel experience. Among respondents with an experience level of 8 or higher, 59.2 % of the Germans were male, while 60.3 % of German respondents with low experience levels were female. Within the New Zealand respondents, there was also a higher self-rating by male respondents of high experience levels (56.0 %), while low experience levels were predominantly selected by female respondents (63.8 %).

There were no statistically significant differences between age groups in the German data set ($p =.73$). Results from the New Zealand data set reveal statistical differences between age groups and indicate a general tendency for the older the respondent the higher the self-rated experience level. While more than half of the respondents with a self-rated low experience level were in the age groups 20–29 years and 30–39 years, 42.2 % of respondents with a reported high experience level were 50 years and older. The results of sociodemographic information by subgroups are summarised in Table 1.

There were also significant differences in scores related to the net household income in both case studies. Results showed a positive relationship between higher income and higher levels of self-rated travel experience. When testing for statistical differences related to the respondents' school qualifications, the test requirements were not satisfied by the German data set (minimum expected cell count less than 1). There were significant differences between respondents with a lower and higher school qualification in the New Zealand data set. While 70.8 % of respondents with a self-rated low experience level indicated that their highest school qualification was either secondary or vocational school, 62.9 % with an experience level of 8 or higher had an undergraduate or postgraduate degree.

Responsible and sustainable travel motives, trip activities and future travel behaviour

In a multiple-choice question, respondents were asked to indicate their major motives to travel for holiday purposes (maximum choice of five answers, percentages expressed in percent of the cases). Among the 14 response options, *open my eyes and broaden my knowledge*, *learn about foreign culture* and *contribute positively to the sustainable development of the destination* were the travel motives with the best fit to the research question presented in this chapter. The travel motive *open my eyes and broaden my knowledge* was selected by 40.8 % of German respondents; the analysis by subgroups showed that the higher the self-rated experience level, the more likely this reason to travel was chosen by German respondents (Figure 2). The *open my eyes and broaden my knowledge* travel reason was also selected by 46.1 % of New Zealand respondents, and although statistically significant differences between subgroups were found, the distribution was slightly different. While *open my eyes and broaden my knowledge* was only selected by a third of respondents with low experience

Table 1: Sociodemographic characteristics by self-rated travel experience level percentages of survey respondents

Characteristic		Germany (n = 500)[a]				New Zealand (n = 500)[a]			
		Total	Low	Medium	High	Total	Low	Medium	High
Gender	Female	50.0	60.3	53.1	40.8	51.8	63.8	51.0	44.0
	Male	50.0	39.7	46.9	59.2	48.2	36.2	49.0	56.0
Age	20–29 years	18.2				20.6	34.3	17.3	16.3
	30–39 years	18.0	not significant			19.6	18.1	21.5	17.0
	40–49 years	25.1	(p > .05)			22.8	21.0	22.7	24.4
	50–59 years	22.0				21.0	19.0	20.4	23.7
	60–69 years	16.8				16.0	7.6	18.1	18.5
Income[b]	Low	22.4	34.8	23.8	14.2	17.7	31.9	16.7	8.1
	Lower middle	42.5	34.8	45.5	41.1	36.2	40.7	37.2	30.6
	Middle	24.4	24.2	22.1	28.4	23.5	17.6	24.7	26.1
	Higher middle	7.5	3.0	6.9	10.6	13.4	8.8	12.1	19.8
	High	3.2	3.0	1.7	5.7	9.1	1.1	9.3	15.3
Education[c]	1	12.4	Test requirements not satisfied (minimum expected cell count > 1)			28.6	47.2	25.5	19.7
	2	33.3				22.3	23.6	25.1	15.9
	3	26.1				24.7	17.0	25.9	28.8
	4	27.5				21.3	5.7	21.2	34.1
	No formal qualification	0.7				3.0	6.6	2.3	1.5

a Sample size for income and education slightly smaller as the questions were not forced. Income: NZ = 417; GER = 438. Education: NZ = 498; GER = 500.
b Prefer not to answer, coded as missing. German questionnaire: Monthly net household income in Euro, converted and annual income projected: Low= Less than NZD 26,500; Lower middle= NZD 26,500 – NZD 52,799; Middle= NZD 52,800 – NZD 79,199; Higher Middle= NZD 79,200 – NZD 105,599; High= NZD 105,600 and more. New Zealand questionnaire: Low= less than NZD 30,000; Lower middle: NZD 30,000 – NZD 70,000; Middle= NZD 70,001 – NZD 100,000; Higher middle: NZD 100,001 – NZD 130,000; High NZD 130,000 and more
c School qualification German questionnaire: 1= Lower secondary education (Hauptschule), 2= Middle secondary education (Realschule); 3= German Abitur; 4= University degree. New Zealand questionnaire: 1= Secondary school, 2= Vocational qualification; 3= Undergraduate degree; 4= Postgraduate degree

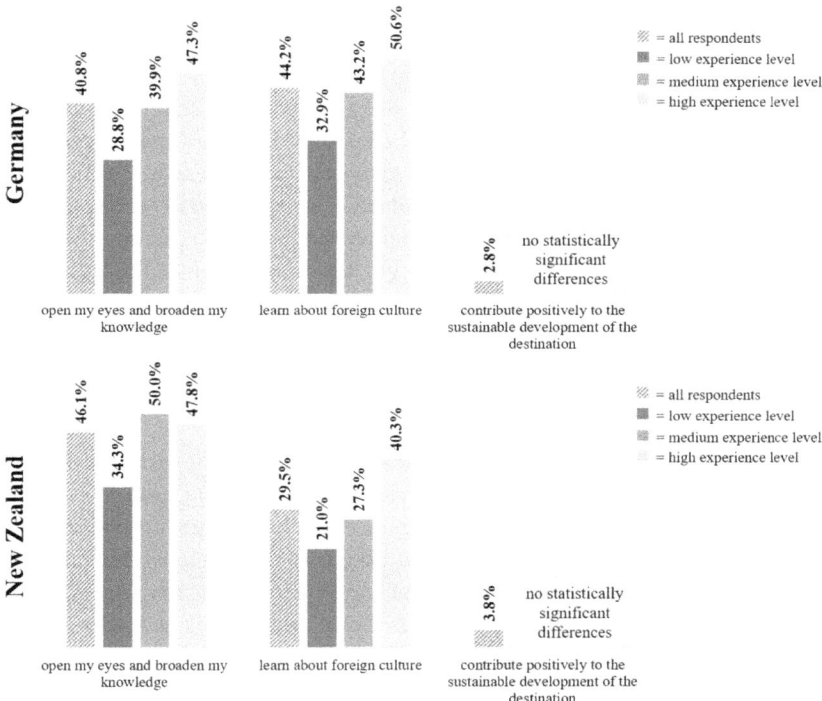

Figure 2: Travel motives of subgroups in Germany and New Zealand (by self-rated travel experience level)

levels (34.3 %), 50.0 % of the respondents with a medium level of experience and 47.8 % of respondents with a high level of experience selected this travel motive.

There were also statistically significant differences between subgroups in relation to the travel motive *learn about foreign culture*. While the motive was selected by 44.2 % of German respondents, the percentage of New Zealanders was considerably smaller (29.5 %). When looking at the approval rates by subgroups, there were greater percentages of respondents with high experience levels who had selected this travel motive. The travel motive *contribute positively to the sustainable development of the destination* was only selected by a considerably smaller amount of German (2.8 %) and New Zealand (3.8 %) respondents, and no statistically significant differences between subgroups were identifiable (Figure 2).

On a scale from 1 (*fully disagree*) to 5 (*fully agree*), respondents were asked to indicate their agreement regarding travel activities undertaken during their most recent international holiday. The trip activities *travelled off the beaten path* and *spent more time in one place to get a real feel* were most strongly associated with responsible and sustainable travel behaviour. Although there were comparably high approval rates for *off the beaten path* travel, Pearson's Chi-square tests were not statistically significant and no differences between respondents with low, medium and high travel experience were found. Regarding the travel activity *spent more time in one place to get a real feel*, statistically significant differences were found only in the German data set. More than two thirds of the respondents with high experience level agreed or fully agreed that they had spent more time in one place to get a real feel, while only 44.1 % of the respondents with low experience levels agreed or fully agreed that they had undertaken this activity during their last international holiday.

Lastly, respondents were asked to indicate their agreement with future travel behaviour on a scale from 1 (*fully disagree*) to 5 (*fully agree*) (Figure 3).

Closely related to the off the beaten path travel behaviour during their last international holiday, respondents were asked to rate their agreement with the statement *want to visit places not many people have been before*. Statistically significant differences were found in both case studies as well as a tendency towards higher approval rates among respondents with higher self-rated experience levels.

Figure 3 reveals a similar tendency yet with greater differences regarding the future intention of *want to experience places like the locals do*. Respondents were also asked to rate their agreement with the statement *want to travel more often in the future*. There was strong agreement with this statement among Germans and the computation of cross-tabulations showed higher approval ratings among respondents with self-rated high experience. While the difference between German respondents with a low experience level (47.9 %) and a high experience level (78.1 %) was comparably large, the difference between respondents with a medium level of experience (61.6 %) was just slightly lower than that of respondents with high levels of experience.

6. Discussion

Findings from both research phases and cases support the assumption of a generally higher level of experience among tourists. Industry representatives from DMOs in Germany and New Zealand agreed that the identification of an experienced tourist requires going beyond previously limited quantifiable factors,

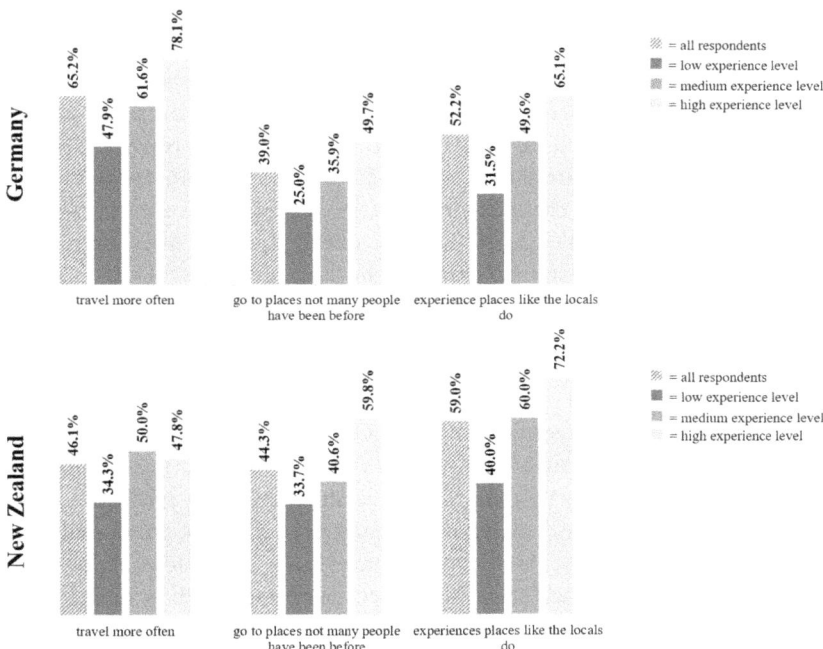

Figure 3: Future travel behaviour of subgroups in Germany and New Zealand (by self-rated travel experience level)

such as age and number of trips (Lehto et al., 2004; Pearce & Caltabiano, 1983; Pearce & Lee, 2005). Instead, it was suggested that an individual's process of experience accumulation and their resulting level of experience is multidimensional and consists of personal identifiers related to the individual's profile and external facilitators related to the external environment. A correlation between higher levels of experience and independent travel forms was proposed and it was further suggested that an experienced tourist's knowledge and awareness positively contribute to responsible and sustainable travel behaviour. This can be linked to active absorption and immersion and thus the education and the escapist realms of experience creation (Pine & Gilmore, 2011).

Despite the experts' general assumption that there is greater awareness of ethical, responsible and sustainable travel choices, the concern was raised about whether this interest is then manifested in actual behaviour, or whether it is a pseudo desire based on socially desirable behaviour. This concern reflects the attitude–behaviour gap identified in past literature (Juvan & Dolnicar, 2014; Lee et al., 2017). Like previous studies, industry representatives referred to the unwillingness to pay higher prices for sustainable offers, yet also admitted to a lack of sustainability information and education throughout the service chain. However, it was also noted that tourists with higher levels of travel experience prefer travel forms that are more inclined to contribute positively to the sustainable development of the tourism industry. As these travel forms are not necessarily directly labelled as sustainable offers, they are often overlooked when investigating the sustainable travel behaviour of today's tourists.

Results from the quantitative online survey of 500 German and 500 New Zealand tourists support the industry experts' assumption that there is a higher likelihood that tourists with higher experience levels choose more sustainable travel forms. With respect to general travel motives and future travel intentions, there were significant differences between subgroups with an overall tendency that the higher the experience level, the higher the approval ratings of respondents. The results also show that tourism offers that are directly labelled as being sustainable are not necessarily demanded and preferentially purchased. This adverse effect has been discussed in previous research (Banfield et al., 2014). At the same time, results clearly show that experienced tourists show high interest for activities and experiences that positively contribute to the sustainable development of the tourism industry, yet are not directly understood as sustainable travel forms but are often outside of the actual tourism bubble (e.g., unstaged engagement with local communities). Thus, there is potential for the tourism industry to develop more sustainably by advocating experiences that engage with local communities and encourage dispersion.

However, there were no statistically significant differences found between tourists with different self-assessed experience levels regarding the actual activities undertaken during their last international holiday. It can be concluded that the differing intentions and planned behaviour of tourists with different experience levels do not necessarily translate into differences in actual behaviour. This further supports the already discussed attitude–behaviour gap and leads to the postulation that this behaviour gap exists across different segments of tourists.

The higher than average approval ratings of experienced tourists for the future holiday intention of *travel more often* also puts into question whether their general awareness and sustainable intentions will translate into actual sustainable travel behaviour. With the ever-increasing numbers of international tourists and growing use of transport options, current travel behaviour and general mobility trends have been classified as unsustainable. Although more experienced tourists might generally travel more responsibly and sustainably, their overall desire to explore more countries and cultures, and travel further and more often, relativises their tendencies towards more sustainable travel forms. In the past decade, the tourism industry has increasingly endeavoured to achieve more sustainable tourism practices, with governmental support. However, in achieving sustainability goals, the tourism industry must go beyond defining sustainable practices. As shown in the recently published *Sustainable Tourism Programme* by the UNWTO (2018b), a closer relationship between sustainable production and consumption is required to enhance and safeguard the sustainable development of the tourism industry.

7. Conclusion and future research

This study aimed at investigating experienced tourists and their sustainable travel behaviour from a macro-micro perspective by incorporating the views of the supply-side (industry experts) and the demand-side (tourists) of tourism. This approach enabled the investigation of the same research phenomenon from different perspectives and thereby contribute to an understanding of sustainable production (supply-side) and consumption (demand-side). The findings from the mixed-methods research approach demonstrated that there are links between travel motives, travel behaviour and experience levels of tourists. A general tendency for higher experience levels to be associated with more sustainable travel behaviour was found. However, it is questionable whether the experienced tourists' awareness and demand for educational experiences that are more ethically responsible and sustainable translate into actual travel behaviour.

While this study has contributed to shedding more light on the subgroup of experienced tourists, the previously identified challenge of overcoming the attitude–behaviour gap remains unresolved. Despite the endeavours of the tourism industry towards more sustainable tourism practices and governmental support in achieving sustainability goals, the tourism industry still needs a more holistic understanding of contemporary tourists' sustainable intelligence and overall demands. As stated by López-Sánchez and Pulido-Fernández (2016), past research has mostly focussed on the stimulation of sustainable tourism practices from stakeholder perspectives that represent the supply-side, while the demand-side has been widely overlooked. The investigation of behaviour by different subgroups proves to be increasingly difficult as consumers not only behave differently in their home environment compared to being on holiday, tourists are also increasingly hybrid and not only change their behaviour between holiday trips but also during one particular holiday (Boztug et al., 2015; Csikszentmihalyi & Coffey, 2017).

In support of past research, this study has shown that tourists' awareness of and interest in sustainable offers has grown. Today's experienced tourists are not necessarily demanding tourism offers that carry a sustainable label but are seeking experiences outside of the commercial tourism bubble. These experiences, however, support the sustainable development of the tourism industry as this travel behaviour (e.g., off the beaten path or local community engagement) positively contributes to economic, environmental and social well-being. This finding is supported by López-Sánchez and Pulido-Fernández (2016) who suggested that tourists value and interpret sustainability in different ways. To safeguard a destination's sustainable development these different interpretations of sustainability and realisations of sustainable tourism practices need to be addressed. Future research is required in order to understand softer forms of sustainable travel behaviour more comprehensively and to identify segments of the demand that contribute to sustainable tourism consumption.

References

Ajzen, I. (1985). From intentions to actions: A theory of planned behavior. In J. Kuhl & J. Beckmann (Eds.), *Action-control: From cognition to behavior* (pp. 11–39). Heidelberg, Germany: Springer.

Alreck, P. L., & Settle, R. B. (2004). *The survey research handbook* (3 ed.). Boston, MS: McGraw-Hill/Irwin.

Antimova, R., Nawijn, J., & Peeters, P. (2012). The awareness/attitude-gap in sustainable tourism: A theoretical perspective. *Tourism Review, 67*, 7–16. https://doi-org.ezproxy.aut.ac.nz/10.1108/16605371211259795

Banfield, J. C., Shepherd, S., & Kay, C. (2014). Consequences of system defense motivation for individuals' willingness to act sustainably. In H. C. M. van Trijp (Ed.), *Encouraging sustainable behavior: Psychology and the environment* (pp. 111–123). New York, NY: Psychology Press.

Bosangit, C., Hibbert, S., & McCabe, S. (2015). "If I was going to die I should at least be having fun": Travel blogs, meaning and tourist experience. *Annals of Tourism Research, 55*, 1–14. https://doi.org/10.1016/j.annals.2015.08.001

Boztug, Y., Babakhani, N., Dolnicar, S., & Laesser, C. (2015). The hybrid tourist. *Annals of Tourism Research, 54*, 190–203. https://doi.org/10.1016/j.annals.2015.07.006

Braun, V., & Clarke, V. (2006). Using thematic analysis in psychology. *Qualitative Research in Psychology, 3*(2), 77–101. http://dx.doi.org/10.1191/1478088706qp063oa

Chang, S. (2018). Experience economy in hospitality and tourism: Gain and loss values for service and experience. *Tourism Management, 64*, 55–63. https://doi.org/10.1016/j.tourman.2017.08.004

Chen, J. S., Prebensen, N. K., & Uysal, M. (2014). Dynamic drivers of tourist experiences. In N. K. Prebensen, J. S. Chen, & M. Uysal (Eds.), *Creating experience value in tourism* (pp. 11–21). Wallingford Oxfordshire, England: CABI.

Creswell, J. W., & Plano Clark, V. L. (2011). *Designing and conducting mixed methods research* (2 ed.). Los Angeles, CA: SAGE.

Csikszentmihalyi, M., & Coffey, J. (2017). Why do we travel? A positive psychological model for travel motivation. In S. Filep, J. Laing, & M. Csikszentmihalyi (Eds.), *Positive tourism* (pp. 122–132). New York, NY: Routledge.

Denscombe, M. (2017). *The good research guide: For small-scale social research projects* (6 ed.). London, England: Open University Press.

Ek, R., Larsen, J., Hornskov, S. B., & Mansfeldt, O. K. (2008). A dynamic framework of tourist experiences: Space-time and performances in the experience economy. *Scandinavian Journal of Hospitality and Tourism, 8*, 122–140. https://doi.org/10.1080/15022250802110091

Fennell, D. A. (2006). *Tourism ethics*. Clevedon, England: Channel View Publication.

Gössling, S., & Peeters, P. (2007). 'It does not harm the environment!' An analysis of industry discourses on tourism, air travel and the environment. *Journal of Sustainable Tourism, 15*(4), 402–417. https://doi.org/10.2167/jost672.0

Hall, C. M. (2013). Framing behavioural approaches to understanding and governing sustainable tourism consumption: beyond neoliberalism, "nudging" and "green growth"? *Journal of Sustainable Tourism, 21*(7), 1091–1109. https://doi.org/10.1080/09669582.2013.815764

Holden, A. (2016). *Environment and tourism* (3 ed.). London, England: Routledge.

Jantzen, C. (2013). Experiencing and experiences: A psychological framework. In J. Sundbo & F. Sørensen (Eds.), *Handbook on the experience economy* (pp. 146–170). Cheltenham, England: Edward Elgar Publishing.

Juvan, E., & Dolnicar, S. (2014). The attitude–behaviour gap in sustainable tourism. *Annals of Tourism Research, 48*, 76–95. https://doi.org/10.1016/j.annals.2014.05.012

Kruger, M., & Saayman, M. (2017). An experience-based typology for natural event tourists. *International Journal of Tourism Research, 19*(5), 605–617. https://doi.org/10.1002/jtr.2133

Lanzini, P. (2018). *Responsible citizens and sustainable consumer behavior*. London, England: Routledge.

Lee, H. Y., Bonn, M. A., Reid, E. L., & Kim, W. G. (2017). Differences in tourist ethical judgment and responsible tourism intention: An ethical scenario approach. *Tourism Management, 60,* 298–307. https://doi.org/10.1016/j.tourman.2016.12.003

Lehto, X. Y., O'Leary, J. T., & Morrison, A. M. (2004). The effect of prior experience on vacation behavior. *Annals of Tourism Research, 31,* 801–818. https://doi.org/10.1016/j.annals.2004.02.006

López-Sánchez, Y., & Pulido-Fernández, J. I. (2016). In search of the pro-sustainable tourist: A segmentation based on the tourist "sustainable intelligence". *Tourism Management Perspectives, 17,* 59–71. https://doi.org/10.1016/j.tmp.2015.12.003

Lück, M. (2002). Looking into the future of ecotourism and sustainable tourism. *Current Issues in Tourism, 5*(3&4), 371–374. https://doi-org.ezproxy.aut.ac.nz/10.1080/13683500208667930

Margetts, E. A., & Kashima, Y. (2017). Spillover between pro-environmental behaviours: The role of resources and perceived similarity. *Journal of Environmental Psychology, 49,* 30–42. https://doi.org/10.1016/j.jenvp.2016.07.005

Mossberg, L. (2007). A marekting approach of the tourist experience. *Scandinavian Journal of Hospitality and Tourism, 7,* 59–74. https://doi.org/10.1080/15022250701231915

Muler Gonzalez, V., Coromina, L., & Galí, N. (2018). Overtourism: Residents' perceptions of tourism impact as an indicator of resident social carrying capacity - case study of a Spanish heritage town. *Tourism Review.* https://doi.org/10.1108/TR-08-2017-0138

Oh, H., Fiore, A. M., & Jeoung, M. (2007). Measuring experience economy concepts: Tourism applications. *Journal of Travel Research, 46,* 119–132. https://doi.org/10.1177/0047287507304039

Pearce, P. L., & Caltabiano, M. L. (1983). Inferring travel motivation from travelers' experiences. *Journal of Travel Research, 22*(2), 16–20. https://doi.org/10.1177/004728758302200203

Pearce, P. L., & Lee, U.-I. (2005). Developing the travel career approach to tourist motivation. *Journal of Travel Research, 43,* 226–237. https://doi.org/10.1177/0047287504272020

Pine, B. J., & Gilmore, J. H. (1999). *The experience economy: Work is theatre & every business a stage*. Boston, MA: Harvard Business School Press.

Pine, B. J., & Gilmore, J. H. (2011). *The experience economy*. Boston, MA: Harvard Business Press.

Postma, A., & Schmuecker, D. (2017). Understanding and overcoming negative impacts of tourism in city destinations: Conceptual model and strategic framework. *Journal of Tourism Futures, 3*(2), 144–156. https://doi.org/10.1108/JTF-04-2017-0022

Prillwitz, J., & Barr, S. (2011). Moving towards sustainability? Mobility styles, attitudes and individual travel behaviour. *Journal of Transport Geography, 19*(6), 1590–1600. https://doi.org/10.1016/j.jtrangeo.2011.06.011

Quan, S., & Wang, N. (2004). Towards a structural model of the tourist experience: An illustration from food experiences in tourism. *Tourism Management, 25*, 297–305. https://doi.org/10.1016/S0261-5177(03)00130-4

Seeler, S., Lück, M., & Schänzel, H. A. (2018). The concept of experience in tourism research: A review of the literature. In A. Ali & J. Hull (Eds.), *Multi-stakeholder perspectives of the tourism experience: Responses from the International Competence Network of Tourism Research and Education (ICNT)* (pp. 203–226). Frankfurt am Main, Germany: Peter Lang.

Sharifpour, M., & Walters, G. (2014). The interplay between prior knowledge perceived risk and the tourism consumer decision process: A conceptual framework. *The Marketing Review, 14*, 279–296. https://doi.org/10.1362/146934714X14024779061956

Thøgersen, J., & Ölander, F. (2003). Spillover of environment-friendly consumer behaviour. *Journal of Environmental Psychology, 23*, 225–236. https://doi.org/10.1016/S0272-4944(03)00018-5

Tung, V. W. S., & Ritchie, J. R. B. (2011). Exploring the essence of memorable tourism experiences. *Annals of Tourism Research, 38*, 1367–1386. https://doi.org/10.1016/j.annals.2011.03.009

United Nations Environment Programme, & World Tourism Organization. (2005). Making tourism more sustainable: A guide for policy makers. Retrieved from http://www.unep.fr/shared/publications/pdf/dtix0592xpa-tourismpolicyen.pdf

van Trijp, H. C. M. (2014). Introduction. In H. C. M. van Trijp (Ed.), *Encouraging sustainable behavior: Psychology and the environment* (pp. 1–10). New York, NY: Psychology Press.

World Tourism Organization. (2018a). 2017 International tourism results: The highest in seven years. Retrieved from http://media.unwto.org/press-release/2018-01-15/2017-international-tourism-results-highest-seven-years

World Tourism Organization. (2018b). The Sustainable Tourism Programme of the 10-Year Framework of Programmes on Sustainable Consumption and

Production Patterns. Retrieved from http://cf.cdn.unwto.org/sites/all/files/docpdf/brochure10yfpstpenupdated17oct2016.pdf

World Tourism Organization, & United Nations Development Programme. (2017). Tourism and the Sustainable Development Goals - Journey to 2030. Retrieved from https://www.e-unwto.org/doi/pdf/10.18111/9789284419401

World Travel & Tourism Council. (2017). Travel & tourism: Global economic impact & issues 2017. Retrieved from https://www.wttc.org/-/media/files/reports/economic-impact-research/2017-documents/global-economic-impact-and-issues-2017.pdf

Marit Gundersen Engeset and Jan Velvin

Identifying travel motives for visitors to Hemsedal during low season

1. Introduction

Hemsedal is a skiing resort in the south of Norway. It is the second largest ski area in Norway and a part of the second biggest ski company in the world, Skistar AB, and tourism has grown to be the main industry in the municipality. Hemsedal is also one of few rural municipalities in Norway, which experience population growth. There are around 6,000 commercial beds in the destination, with 550,000 commercial overnight stays in the winter season. In addition, overnight stays in private cabins amount to approximately 300,000 (Velvin & Kvikstad, 2010). The figures for the summer season are more uncertain, at approximately around 100,000 commercial nights. There has been a steady increase in the number of guests from 2011 to 2018; mainly in winter but the summer, market is also picking up.

The seasonality of the skiing product is a challenge for destinations like Hemsedal (Pegg et al., 2012, Engeset & Velvin, 2016): Over-capacity, non-utilization of infrastructure, and reduction in the workforce during low season create unemployment and loss of services for the local community. Businesses experience problems with employee retention and profitability. Visitors to the destination may experience overcrowding and high demand on services during high season and a reduction in services during low season. Dedicated to creating stable and profitable, workplaces and a high level of service to tourists throughout the year, Hemsedal has been working to increase their low-season demand over the past decade. Building on their reputation as an energetic, innovative skiing resort for families and young adults, they have focused on elements related to nature-based activities targeting the same market segments during summer. The main activities in the summer season are hiking and biking in the mountains. Other popular activities are trout safaris (river snorkelling) and fly-fishing. Hemsedal's main competitive advantages in the summer season are the mountain scenery and the close proximity to the popular Norwegian fjords.

The strategy to target the same market segments in low season as in high season, with activities that address the same types of needs (nature-based activities) has its obvious advantages in that The Hemsedal Tourist office already know the target market very well. The target market is familiar with the Hemsedal brand, and the destination has access to communication channels to reach them. On the other hand, communicating a summer product to market segments that associate the destination with winter tourism may be difficult. The key to success is to identify how Hemsedal can capitalize on their winter destination reputation while at the same time communicating attractiveness of their summer products. This requires detailed insight into what drives tourists' demand during summer. Therefore, to be able to develop their summer product further and to communicate its attractiveness to their target market, it is important for the destination to understand what motivates their summer visitors to choose Hemsedal. The purpose of this chapter is to identify those motives and to evaluate Hemsedals' summer products by investigating the level of success in addressing their visitors' motivations to visit the destination.

The chapter is organised as follows: First, we review literature on tourist motivation. Second, the results of a survey testing the motivations, satisfaction, and future visit intentions of visitors to Hemsedal during the 2015 summer season are presented and analysed. Then, we discuss the implications of motivation theory and the results of the survey for Hemsedal has continued effort to strengthen their summer products. Finally, we discuss how this article is relevant for the development of other destinations with high seasonal fluctuation in demand.

2. Tourist motivations

A widely accepted general definition of a motive is "an internal factor that arouses, directs, and integrates a person's behaviour" (Murray, 1964, p.7). Tourism motivation is defined in several ways in the literature: "a meaningful state of mind which adequately disposes an actor or group of actors to travel, and which is subsequently interpretable by others as valid explanations of such decisions" (Dann, 1981, p.211), or "a dynamic process of internal psychological factors (needs and wants) that generate a state of tension or disequilibrium within individuals" (Crompton & McKay, 1997, p.427). More simply stated, tourist motivations are reasons for travel. They help explain why tourists travel in the first place, and why they choose one destination over another. These reasons are complex, and a number of considerations and desires interplay to create an overall motivational state leading to a certain choice. For example, a mother may balance her own needs to rest and get a break from a busy life with the desire to see a particular site and the needs of her family with practical considerations that are constraining their options for vacation. In tourism research, this complexity is reduced by identifying motivational factors as either internal or external (Gnoth, 1997). Internal factors are individual desires and needs, such as the need for relaxation, adventure, family togetherness, prestige, escape, health, and excitement (Crompton, 1979). External factors are connected to situational factors and destination attributes (Uysal & Hagan, 1993, Yoon & Uysal, 2005). Tourism literature refers to internal factors as push motivations – tourists are "pushed" towards particular travel choices by their own internal desires and needs and "pulled" towards a choice by particular characteristics of the travel destinations (Caber & Albayrak, 2016; Cha, McCleary, & Uysal, 1995; Oh, Uysal, & Weaver, 1995; Yoon & Uysal, 2005).

The literature provides many different types of push and pull motivations. For example, in a study of tourists visiting Northern Cyprus, Yoon and Uysal (2005) identified eight push motivation factors including, for example, excitement, achievement, and family togetherness, and ten different pull motivation factors such as, modern atmosphere and activities, natural scenery, and different culture. Understanding the types of push and pull motivations that drive tourists' decisions to visit a destination is important. Insight into the push motivations can help destinations develop attractive products addressing the most important motives that drives demand, as well as identifying important aspects to focus on in communication with the target market. Identifying pull motives shows what aspects of the destination drives tourist demand. This knowledge is an important input to the development of targeted communication strategies for the future.

3. Tourist satisfaction and behavioural intentions

Satisfaction is an important aspect of tourists' loyalty, and likely leads to revisits and positive word of mouth referrals (Gundersen, Heide, & Olsson 1996). One of the most widely accepted model for customer satisfaction is the expectancy disconfirmation model first proposed by Oliver (1980). According to this model, customers form expectations about product experiences based on their own needs and motivations, marketing communication, publicity in media, and recommendations from friends and family. After experiencing a product, the actual product performance is compared to expected performance. If expectations are met or exceeded, the customer will be satisfied. If there is a negative gap between expectations and performance, the customer will be dissatisfied. Push and pull motivations are important for ensuring visitor satisfaction due to their role in shaping expectations. In their study of visitors to Northern Cyprus,

Yoon & Uysal (2005) found that push motivations have a positive direct and indirect effect on destination revisit intentions and a positive direct effect on satisfaction. Surprisingly, they also found that pull motivations had a negative effect on satisfaction. This negative effect may be attributed to a lack of ability of the destination to meet tourists' expectations towards the attributes that were the most important motivational factors in their decision making process, or to other factors relating to measurement or model estimation errors. Regardless of this, their research clearly showed that push and pull motives are indeed important antecedents of travel satisfaction and revisit intentions. Understanding the relative importance of different types of push and pull motivations for predicting satisfaction and behavioural intentions at a particular destination will therefore give important insight into the drivers of demand for that destination.

The literature reviewed suggest that insight into motivational factors driving the demand for a destination is important for tourist destinations. First, such insight can be used to identify areas of highest importance for tourist choice, information that can be used to develop competitive offerings and identify how to communicate the offerings to the target market. Second, such insight can help understanding the relative strength of the impact of the different types of motivational drives on satisfaction and revisit intentions. This knowledge is important to gain a deeper understanding of tourists' expectations and satisfaction. The purpose of this study was to provide this insight for Hemsedal. More specifically, it aimed (1) to identify different push and pull motivational factors driving demand for Hemsedal's low season (May-October) products, (2) assess the relative importance of each factor on the choice to visit Hemsedal during the low season, and (3) analyse the relative impact of each factor on different measures of satisfaction and behavioural intentions.

4. Methods: Questionnaire development and data collection

There are no established scales that can be used to measure tourist motivation. Given that, each destination is unique such a scale would have to consist of a large number of items to cover all possible aspects for any destination. Hence, this study had to rely on a number of different sources when generating items for the questionnaire (Chan & Baum 2007, Crompton & McKay 1997, Huang & Hsu 2009, Yoon & Uysal 2005). In addition, managers, employees, and tourists visiting Hemsedal were interviewed to generate items for measuring push and pull travel motivations. Specifically, tourists were approached and asked if they could list (1) the personal "experiences" they had hoped to achieve, and

(2) features of Hemsedal that seemed particularly attractive to them when choosing to spend time in Hemsedal. Similarly, managers and employees were asked to list what they thought would be important personal factors and aspects of the destination that tourists had emphasised when choosing to come to Hemsedal. The responses had many similarities across sources; items mentioned by tourists were also mentioned by employees and managers, and they were consistent with many of the items used in previous studies. Sorting the different aspects mentioned removing overlapping items and adding items from literature that had not been mentioned, resulted in a list of 40 items measuring push motivations and 30 pull motivation items. These items were measured on seven point Likert-type scales ranging from "not at all important" to "very important".

Eight questions were developed to capture the visitors' overall satisfaction and behavioural intentions. The questions were measured on a seven point Likert-type scale ranging from "do not agree at all" to "strongly agree". Two questions measured the overall satisfaction: "overall, I am very satisfied with my stay in Hemsedal" and "my visit to Hemsedal gave me high value for money". Behavioural intentions were measured by two word of mouth (WOM) items: "I will share positive information about Hemsedal in social media" and "I will share positive information about Hemsedal with friends and family"; three measures of revisit intentions during low season – "I wish to return to Hemsedal during spring/summer/fall the next 2–3 years" and one measure of revisit intentions during high (winter) season.

Data were collected among visitors to Hemsedal during the low season (May-October) in 2015. Tourists were approached and asked if they would be willing to complete an online questionnaire measuring their experience in Hemsedal. Those that agreed received a link to the questionnaire by email after returning home. A total of 320 usable responses were received.

5. Results

Demographic profile of the respondents

A total of 149 (46.6 %) of the respondents were male. The respondents were quite evenly distributed across age groups with the majority (28 %) belonging in the 40–50 years age bracket. A total of 58.8 % of the respondents were Norwegian, the rest were mainly from the Netherlands (16 %) and Denmark (11.6 %). Only 1.5 % were from countries outside Europe. Less than half of the respondents (43.2 %) were in the median income group (€30,000 – 50,000 per annum). Some of the respondents travelled privately with friends (23.2 %) and/or families with children (54.5 %), or without children (32.9 %). For those traveling with children, the majority had children over the age of seven (96.5 %). Rented cabins or apartments were the most prominent type of accommodation (55.9 %), and many respondents (43.9 %) spent more than five nights at the destination. Most of the respondents (90.3 %) came to Hemsedal using their private car. This profile of the respondents matches the target market of Hemsedal as a winter destination: relatively young people with average income travelling with friends or family by car and renting a cabin or apartment during their stay.

Identifying push and pull motivational factors

The forty items measuring push motivations were submitted to a principal component analysis with varimax rotation to identify the underlying factors. The analysis resulted in an eleven-factor solution with eigenvalues greater than 1 (variance explained = 68 %). The factor solution was inspected and items with small (<.5) factor loadings removed. The twenty-nine remaining items were submitted to a new principal component analysis and this time the solution suggested nine factors (variance explained = 70 %). The factors were labelled according to the meaning derived from the item wordings. The nine factors with factor loadings are presented in Table 1.

Table 1: Factor solution and composite means for push motivations

Factor Items	1	2	3	4	5	6	7	8	9
Adventure									
Be physically active	**.75**	.21	.15	-.19	-.19	.00	.00	.13	-.01
Experience excitement	**.75**	-.06	.10	.16	.26	.05	.06	.06	-.03
Challenge myself	**.74**	.15	.00	.16	.17	.20	-.01	.02	.16
Satisfy my need for adventure	**.66**	.15	.12	.37	.09	.04	.07	-.01	.08
Sense of freedom									
Be free to do what I want	.18	**.56**	-.04	.32	-.02	.15	.16	.00	-.15
Get a sense of freedom	.29	**.59**	.13	.37	.03	.08	.16	.02	.05
Get back in touch with myself	.15	**.62**	-.06	.06	.43	.15	-.06	.09	-.07
Experience quietness	-.04	**.79**	.11	.10	-.13	.01	.14	.02	.16
Experience serenity	.07	**.81**	.16	-.05	.01	.16	.05	.01	.20
Gather strength	.08	**.62**	.17	-.02	.15	.46	.02	-.01	.27

Table 1: Continued

Factor Items	1	2	3	4	5	6	7	8	9
Novel experiences									
Try new food	.10	.02	.04	.17	**.71**	-.08	.03	.24	.00
Experience a new way of life	.16	.10	.07	.24	**.74**	.0	-.08	.02	.00
Get a sense of luxury	.15	-.15	.08	.09	**.54**	.12	.36	.00	.29
Follow traditions									
Do what I usually do on vacation	.06	.09	.02	.07	.00	.00	**.90**	.06	.12
Stick with my vacation traditions	.01	.16	-.04	.11	.04	.00	**.88**	-.01	.11
Prestige									
Go places none of my friends have been	-.07	.03	.00	**.66**	.30	.13	.22	.04	-.06
Talk about the vacation when I get home	.00	.19	.12	**.59**	.28	.21	.08	.11	.23
See as much as possible	.21	.11	-.01	**.75**	.06	.00	-.01	-.01	.22
Explore new things	.46	.13	.10	**.65**	.09	.11	.00	.06	.10
Sense of togetherness									
Experience community	.12	.09	**.90**	.01	.07	.10	-.01	-.01	.23
Share experiences	.14	.08	**89**	.08	.03	.06	-.01	.05	.22
Get a sense of belonging	.03	.11	**.87**	.02	.05	.08	-.02	-.04	.01
Break from everyday life									
Get a break from a busy job	.10	.16	.07	.03	.12	**.81**	.03	-.01	.03
Be with family	.03	.01	.30	.07	-.22	**.52**	-.03	.28	.08
Get away from the demands in my everyday life	.11	.29	.04	.20	.05	**.78**	.01	-.04	.01
Roots									
Visit places my family comes from	.06	.03	.01	.09	.12	.08	.07	**.87**	.02
Visit friends and relatives	.01	.12	.07	-.02	.11	-.03	.05	**.88**	.09
Low risk									
Be in control	.14	.28	.20	.20	.11	.03	.14	-.02	**.70**
Feel certain that I will experience what I want	.16	-.05	-.10	.14	-.02	.08	.17	.15	**.76**

The 30 pull motivation items were submitted to a principal component analysis with varimax rotation. Two items with loadings below .5 were removed. The 28 remaining items were analysed again, and the final solution suggested seven underlying factors with eigenvalues above 1 (variance explained = 67 %). Table 2 shows the factor solution.

Table 2: Factor solution for pull motivations

Factor Items	1	2	3	4	5	6	7
Nature							
Plenty of outdoor space	**.58**	.11	.12	-.04	.02	-.20	.43
Beautiful scenery	**.82**	.05	.05	-.03	.12	-.13	.12
Many hiking opportunities	**.78**	-.01	.06	.10	.09	.06	-.09
Quiet surroundings	**.57**	.08	.09	-.37	.18	-.03	.32
Close to nature	**.88**	.02	-.05	-.06	.01	.05	.06
Supports a sporty lifestyle	**.71**	.10	.05	.15	.26	.24	-.05
Clean environment	**.73**	.16	.09	.00	.08	.06	.25
Many opportunities for experiences in nature	**.75**	.09	.03	.00	.09	.26	-.16
Culture							
Historic places	.16	**.65**	.28	.04	-.08	-.23	.23
Many opportunities for cultural experiences	.25	**.66**	.37	.14	-.06	.11	-.03
Guided tours	.05	**.71**	.06	.17	.11	.36	-.03
Sightseeing	.00	**.77**	.02	.09	.22	.07	.16
Museums	.03	**.77**	.13	.12	-.03	-.08	.14
Restaurants and shopping							
High quality restaurants	.01	.13	**.74**	.33	.23	.16	.02
High variety of restaurants	.11	.15	**.65**	.05	-.01	-.02	.31
Local food	.08	.33	**.70**	-.01	.15	.12	-.03
Many opportunities for shopping	.02	.08	**.79**	.30	.20	.24	.05
Events							
Festivals and concerts	-.08	.27	.24	**.66**	.08	.17	.09
Lively nightlife	-.06	.08	.27	**.75**	.03	.11	.06
Sports events	.02	.35	.04	**.58**	.32	.02	.32
Activities and attractions							
Many activities	.31	.09	.11	.31	**.68**	.15	.11
Many opportunities for sporting activities	.34	-.02	.03	.28	**.65**	.25	-.12
Many attractions	.27	.45	.22	.15	**.54**	-.11	.00
Children friendly	.03	.01	.20	-.06	**.75**	.02	.09
Safety							
Low cost accommodation	-.04	.14	.04	.12	-.01	.06	**.75**
High personal safety	.32	.14	.17	.10	.12	.09	**.65**

Table 2: Continued

Factor Items	1	2	3	4	5	6	7
Convenience							
Easy to get there	.19	.10	.26	-.03	.20	**.65**	.30
Close to home	.10	-.03	.16	.18	.01	**.77**	-.04

The resulting factor solution for motivations to visit Hemsedal corresponds well to previous research on push and pull motivation (Crompton, 1979, Uysal & Hagan 1993, Yoon & Uysal 2005).

Satisfaction and behavioural intentions

The questions measuring satisfaction (Cronbach's alpha =.66), WOM (Cronbach's alpha=.59), and Low season revisit intentions (Cronbach's alpha = .78) were averaged to form one measure for each of the three constructs. Results showed that the summer guests at Hemsedal were very satisfied (M=5.95, SD=1.17), and the intention to recommend (M=5.71, SD=1.41) and return during low season the next two-three years (M=5.20, SD=1.63) were also very high. In addition, the respondents reported a high intention to return during winter season as well (M=5.11, SD=2.14), indicating that a satisfying experience during low season may lead to intentions to return in high season.

Visitors' motivation and impact on satisfaction and behavioural intentions

To identify what types of push and pull motivations visitors to Hemsedal perceived as most important when choosing to visit the destination, the composite means for each of the dimensions presented in Tables 1 and 2 were calculated. In addition, to understand the impact of push, pull motivations on satisfaction and behavioural intentions for Hemsedal, a correlation analysis between the satisfaction and behavioural intention variables and the push, and pull motivational factors was performed. The results for this analysis are presented in Table 3.

Table 3: Composite means and correlation coefficients

		Correlation coefficients			
	Composite mean	Satisfaction	WOM	Revisit intentions low season	Revisit intentions high season
Push Motivations					
Adventure	4.56	.25**	.32**	.21**	.25**
Sense of freedom	4.90	.20**	.15*	.12*	-.07
Novel experiences	3.29	.12*	.04	.19**	.19**
Follow traditions	3.59	.12*	.00	.09	.04
Prestige	3.93	.15**	.08	.04	.02
Sense of togetherness	5.74	.20**	.23**	.20**	.15*
Break from everyday life	5.38	.21**	.28**	.14*	.14*
Roots	2.24	.08	.03	.20**	.13*
Low risk	4.24	.10	.03	.02	-.03
Pull Motivations					
Nature	5.59	.35**	.24**	.29**	.15*
Culture	2.36	.10	.04	.20	.01
Restaurants and shopping	3.27	.11	.12	.24**	.15*
Events	1.99	-.02	.04	.18**	.26**
Activities and attractions	4.17	.25**	.24**	.31**	.41**
Safety	4.22	.16**	.03	.02	-.04
Convenience	3.87	.08	.07	.16**	.21**

Note: *: $p < .05$; **: $p > .01$.

Table 3 indicates a high correspondence between what the guests report as most important factors (represented by the composite means), and the strength of the correlation between each motivational factor and satisfaction and behavioural intentions. Internal drives relating to adventure seeking, need for freedom, need for togetherness, and desire to get a break from everyday life are important push motivational factors. Nature as well as activities and attractions are important pull motivational factors.

To get a better understanding of the relationships, four mediation analyses we performed, using the two most important push factors (sense of togetherness, break from everyday life) and pull factors (nature, activities and attractions) as independent variables regressed on revisit intentions during low season with satisfaction as mediator. We tested the significance of the proposed effects using mediation analysis with bootstrapping procedures (Hayes, 2013). The results of these analyses are presented in Figure 1.

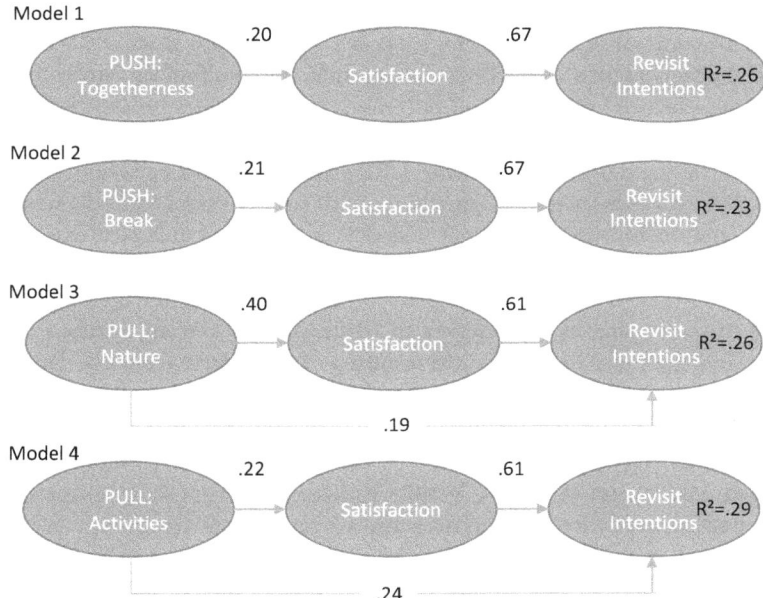

Figure 1: Results from mediation analyses

Unstandardised indirect effects were computed for each of 10,000 bootstrapped samples, and the 95 % confidence interval ranged from .05 to .23 for Model 1, .07 to.22 for Model 2, .15 to .36 for Model 3, and .07 to .20 for Model 4. Thus, the indirect effect was statistically significant for all four models. All path coefficients were significant at $p < .001$. The pull motivation factors had significant direct and indirect effect on revisit intentions, while the push motivation factors had only indirect effect. These results support the notion that motivation has significant effect on revisit intentions, and that satisfaction mediates this effect.

6. Discussion and Conclusion

This study showed that a number of different types of push and pull motivations are important for guests visiting Hemsedal during summer season. The motivation factors identified correspond to previous research and support the notion that visitors are attracted by a combination of their own needs and the characteristics of the destination. Support for the importance of push and pull motivations in predicting satisfaction and behavioural intentions was also found. In contrast to results from Yoon & Uysal's (2005) study, who found a negative effect of pull motivation on satisfaction, the results of this study showed a positive effect of both push and pull motivational factors on satisfaction. In addition, it is demonstrated that these types of push and pull motivations may differ in the strength of relationship with satisfaction and behavioural intentions. This research therefore contributes with a more nuanced understanding of the relationships between push and pull motivations and post-experience evaluations.

The study also has important practical implications for managers of destinations. It shows that by considering the different types of motivations, managers can get a more detailed understanding of what drives tourists' choice to visit their destination and to return. Further, a number of items are presented that can be adopted and used by other destinations in future studies.

Finally, the research can be used by Hemsedal when developing strategies for their low season products. Results indicate that Hemsedal's visitors are motivated by needs for togetherness and to get a break from their everyday lives when they visit Hemsedal. Correlation and regression analyses show that people who are driven by those needs are also likely to be satisfied and to return to Hemsedal on another occasion. Communicating the destination's ability to meet those needs and developing products that facilitate the satisfaction of such needs are viable strategies for Hemsedal in their efforts to position the destination as a place to visit during summer season. This can be done by offering activities that can be enjoyed by a family or group of friends together, and by ensuring that the holidays will be mostly risk-free. In terms of pull factors, Hemsedal should keep capitalising on its reputation as a nature-based destination. This image can be strengthened by developing attractive outdoor activities. Such image corresponds well with the destination image built around the high season product and should be easy to communicate to people that are already aware of the destination image.

References

Caber, M. & Albayrak, T. (2016). Push or pull? Identifying rock climbing tourists' motivations, *Tourism Management*, 55, 74–84.

Cha, S., McCleary K., & Uysal M. (1995). Travel motivation of Japanese overseas travelers: A factor-cluster segmentation approach. *Journal of Travel Research*, 34(1), 33–39.

Chan, J. K. L., & Baum, T. (2007). Motivation factors of ecotourists in ecolodge accommodation: The push and pull factors, *Asia Pacific Journal of Tourism Research*, 12(4), 349–362.

Crompton, J.L. (1979), "Motivations of pleasure vacation," *Annals of Tourism Research*, 21 (2), 283–301.

Crompton, J.L., & McKay, S.L. (1997). Motives of visitors attending festival events, *Annals of Tourism Research*, 24(2), 425–439.

Dann, G. M. (1981). Tourist motivations: An appraisal, *Annals of Tourism Research*, 8(2), 189–219.

Engeset, M.G. & Velvin J. (2016),"From winter destination to all-year-round Tourism," in *Mountain Tourism, Experiences, communities, environments, and sustainable futures,* H.Richins & J.Hull (eds) Oxfordshire UK: Cabi, 79–87.

Gnoth, J. (1997), "Tourism motivation and expectation formation," *Annals of Tourism Research*, 6, 408–424.

Gundersen, M., Heide, M., & Olsson U. (1996), "Hotel guest satisfaction among business travellers: What are the important factors?" *Cornell H.R.A. Quarterly*, 37 (2), 72–81.

Hayes, A.F. (2013), *Introduction to Mediation, Moderation, and Conditional Process Analysis: A Regression Based Approach*, Guilford Press, New York, NY.

Huang, S. & Hsu, C.H.C (2009), "Effects of travel motivation, past experience, preeived constraint, and attitude on revisit intentions," *Journal of Travel Research*, 48 (1), 29–44.

Murray E.J. (1964), *Motivation and Emotion*, Englewood Cliffs, NJ: Prentice Hall

Oh, H.C., Uysal, M., & Weaver, P. (1995), "Product bundles and market segmentation based on travel motivations: A canonical correlation approach," *International Journal of Hospitality Management*, 14 (2), 123–137.

Oliver, R.L. (1980), "A cognitive model of the antecedents and consequences of satisfaction decisions," *Journal of Marketing Research*, 17, 46–49.

Pegg, S., Patterson, I.M. & Gariddo, P.V. (2012), "The impact of seasonality on tourism and hospitality operations in the alpine region of New South Wales, Australia," *International Journal of Hospitality Management* 31, 659–666.

Uysal, M. & Hagan, L.R. (1993), "Motivation of pleasure to travel and tourism," in *VNT's Encyclopedia of Hospitality and Tourism*. M.A. Kahn, M.D. Olsen, & T. Var (eds). New York, Van Nostrand Reinhold, 798–810.

Velvin, J., & Kvikstad, T. M. (2010). *Second homes eller den tradisjonelle hytta?: hva genererer det hyttebaserte reiselivet av økonomisk verdiskapning i den enkelte hyttekommune i Buskerud*. (83). Retrieved from Kongsberg: http://idtjeneste.nb.no/URN:NBN:no-bibsys_brage_15700

Yoon Y. & Uysal M. (2005), "An examination of the effects of motivation and satisfaction on destination loyalty: A structural model," *Tourism Management*, 26, 45–56.

Jarmo Ritalahti

Customer insights of sustainability and responsibility in leisure travel intermediation in Finland

1. Introduction

Travel intermediation has changed significantly after the introduction of the Internet and electronic market places supported by it. One of the main questions is that what is the role of traditional travel intermediaries in the dynamic and information rich environment where end-users and service suppliers can connect with each other without the third party (Novak & Schwabe, 2009). According to some authors there will still be a need for traditional travel intermediaries, but they should be able to reposition themselves by providing added value to their customers, and not only acting as ticket reservation or ticketing offices (Dilts & Prough, 2002; Alamdari, 2002; Cheyne, Downes & Legg, 2005). There have been several studies to understand the future of traditional travel intermediaries with high street offices (e.g. Castillo-Manzano & López-Valpuesta, 2010), but how they can remain competitive and viable in filling the changing needs of more sophisticated customers remains a challenge.

Online shopping with sales 24/7 makes the contact with customers and travel service suppliers possible in a virtual environment, and the physical environment is not generally needed anymore. Purchases are online transactions that have also changed the consumer behavior in a radical way. Of course, the evolution has gone through similar steps than with traditional channels including the adaption, acceptance and re-purchasing, but for intermediaries, it is important to understand motives of changing from a traditional intermediary to an online actor. Some of the issues for the traditional intermediaries are related to security, reliability and convenience when in a physical office. For the online agencies, the pros areservice availability 24 hours a day, immediate response time and ease of use (Ritalahti, 2018).

This chapter discusses the sustainability and responsibility of traditional travel agencies and tour operators, how the customers see them, and their importance when choosing an agency. The themes of sustainability are based on the sustainability megatrend and its interpretations. Responsibility rises from customer loyalty and its recent emphasis on employees. The aim of the study is to give agencies responses to understand contemporary customers and their purchasing behavior. The customer data were collected in a survey executed in March and April 2018. It was sent to the member organisations of the Association of Finnish Travel Agents who forwarded it to their customers or people who were in their email databases.

2. Literature review

Travel intermediation is a link in the distribution chain, a dealer between a service provider and its customers (e.g. Lubbe, 2000). Traditionally, travel agencies and tour operators have been intermediaries in the tourism industry. Today, it is difficult to separate a travel agency from a tour operator, but traditionally a travel agency is a retailer and tour operator a wholesaler. Lominé and Edmunds (2007) states that a tour operator pre-arranges and distributes or sells holiday packages to end-users. Now, these packages can also be more tailor made. Travel agencies offer products and services of tour operators, cruise lines, airlines, trains and ferry operators for sale. The main task of a travel intermediary is to bring end-users and service suppliers together and reduce costs between buyers and sellers. Travel intermediaries' importance for suppliers is that they can sell in bulk and, furthermore take a certain risk when selling services to end-users. Tour operators are often tied to contracts with service providers. Consumers used to benefit by avoiding search and transaction costs and getting specific information from agents of tour operators. The purchase of a package commonly offered consumers more reasonable prices. Today, the benefits might not be that obvious. Scaglione and Schegg (2016) state that online travel intermediaries have a lot of market and business knowledge because they collect data of customer behavior. For the consumer, this can mean a reduced choice, higher prices and concentration and consolidation of travel agencies and tour operators under wider brands.

Ritalahti (2018) concludes in his article that traditional travel agencies and tour operators face several challenges. It seems that added-value of a high street travel agency is not related to good service or qualified experts, but to online presence including destination information, booking and purchasing options. The challenges are more the impacts of megatrends on customers. We know that societies of high consumption are getting older, in general countries are more urbanised and better off, and the middle class is growing. Furthermore, technology develops faster and faster. Ritalahti (2018) asserts that megatrends steer more consumers and their purchasing behavior than earlier. If the service supply is available 24/7 online, is there a need for face-to-face contacts any longer? To shop and purchase online became normal and an everyday activity, and is easier and safer than before - consumers' trust has increased.

The megatrend on focus in this chapter is sustainability as a global phenomenon including responsibility. "Megatrends are macroeconomic and geostrategic forces that are shaping the world. They are factual and often backed by verifiable data. By definition, they are big and include some of society's biggest challenges – and opportunities." (PWC 2016)

Dwyer et al. (2009) described megatrends underpinning tourism to 2020 as following: Globalisation and long-term economic trends, social trends, political trends, and environmental trends. Globalisation and long-term economic trends include topics like a growing global economy, globalisation, and breaks to growth. Social trends are listed as population and an ageing society, urbanisation, changing social structures in developed economies, health, values and lifestyles, and education. Political trends include topics such as existing and emerging global players, terrorism, health risks and security, haves versus have nots, political Islam, impact of climate change. Environmental trends are climate change, depletion of natural resources, loss of biodiversity, and other environmental trends.

Oxford Economics (2012) look at macro trends in the travel industry until 2023. Their report focuses more on impacts of mega trends on the industry rather than mega trends themselves. The impacts or trends are the growth of tourism in emerging markets i.e. the increasing visitor flows to Asia Pacific, the Middle East, and Africa. The share of non-OECD countries in global air traffic is expected to grow from 44 % in 2013 to 51 % by 2023. The driver of this change is the economic expansion of these areas, especially in China. The growth of the travel sector in emerging markets also means innovations in service supply to meet the rapidly increasing and changing demand, especially in maturing domestic markets. The use of new technology and experimenting with it is more common in the traditional and advanced markets. Seamless travel based on technological and infrastructural development is growing in importance, and so are demand of services and applications in internet and mobile technologies. For example, the number of mobile device users is growing, especially in emerging markets.

OECD (2017) defines the following megatrends as important with long-term implications (until 2040) for tourism: Globalisation and demographic trends, global shift to low-carbon and resource-efficiency, sustainability, and travel mobility. Globalisation and demographic trends will change visitor demand in tourism. The growing income and education levels in emerging economies also mean an increased importance of these tourism-generating areas. The number of people over 60 years of age is expected to double in the next 35 years. Millennials (generation Y) and generation Z will be the key consumers in tourism by 2040. The global shift to low-carbon resource-efficiency also challenges tourism, but tourism can play a key role in enhancing the shift to more environmentally friendly technology and policy. Mobile technologies change the planning and experiencing of travel. Peer-to-peer supply and the sharing economy using digital platforms will grow in popularity through creating new marketplaces and business models. Transport and its development is essential to the tourism system, by connecting the markets in tourism generating areas to destinations. The topical issues in transport are the growing number of visitors (volume), safety and security concerns, ecological perceptions of tourists, transport innovations, infrastructure, and regulations.

According to Horwath HTL (2017) 10 megatrends influencing mid- and long-term tourism development on the demand side are silver hair tourists, generations Y & Z, a growing middle class, emerging destinations, and political issues and terrorism. The megatrends on the supply side include technological (r)evolution, digital channels, loyalty of customers, health and healthy lifestyle, and sustainability. Silver-haired tourists are a result of ageing population megatrends. The share of people over the age of 60 has increased from 8 % in 1950 to 12 % in 2013 and is expected to reach 21 % by 2050. Seniors, in this case people over 60 years of age, are ready to spend money. They live longer and healthier and are active travellers who are looking for highly personalised services, soft adventure making them feel young, and health and wellness products. Alongside seniors, generations Y and Z will be an important market segment. Their preferred lifestyle means changes in the service supply and supply chain: They use mobile devices, they have a high level of involvement in digital trends, and they are highly educated. Furthermore, they will probably have 17 jobs and 15 places of residence in a lifetime (Horwarth HTL, 2017).

The growing middle class comes from the emerging economies, especially from the Asia-Pacific region. When travelling, the middle class is looking for value for money. They have invested in their own training and are able to speak English, thus the travel supply on the Internet is open to them. The growing middle class and their request for good value for money ervices, opens opportunities to new destinations in these emerging and developing regions. Cities with double digit growth rates are, for example, Istanbul, Seoul, Bangkok, Tokyo, and Taipei. Political issues and terrorism include topics such as governments' ability to tackle the unwanted incidents from terrorism, and how to cope with the increasing number of refugees and illegal immigrants. Transportation security is important for travellers. Technological (r)evolution influences

Table 1. Sustainability as a Megatrend

Sustainability	Increasing energy and water consumption; Rising CO2 emissions; Scarcity of resources; Growing population > increase in food demand; Climate change and extreme weather conditions; Declining biodiversity; Sustainable and smart solutions

tourism by increased demand of mobile services, real time insights and customisation as well as digital interaction, robots in service encounters and wearables.

Travel megatrends of 2017 vary by reports and sources. Skift (2017) defines them as: Human touch, over tourism, artificial intelligence, day tours, ageing population, festivalisation and MICE tourism, new luxury, and dining out. Despite technology increasingly steering most tourism related transactions, people still demand human interactions when travelling. These can be as simple as stories and advices from the locals. Artificial intelligence feeds search engines to help travellers with a good tourism service. It learns how to deliver better results after understanding the customer behavior on the Internet. Today, tourism is enhanced by increasing numbers of flights by budget and network carriers, as well as lower ticket prices. The new growing generations in tourism are post-baby boomers, such as generations X and Y. It is not yet time for generation Z to travel independently. Mass tourism is a phenomenon that has its bright sides, but over tourism is a recent phenomenon where the social carrying capacity of the local people has exceeded its limits. Reasons behind over tourism may be the rapid expansion of low-cost carriers and sharing economy businesses. The nature of businesses, such as Airbnb, has changed in many places from renting one of the empty rooms in one's home to apartments purchased by investors who can realise larger profits with tourists than local people.. Dining out is one of the highlights of the day of a modern traveller. Dining and food are more critical components of the holidays; thus, restaurants have turned even to destinations itself. Also, in food, the local touch is most important.

According to the managing director of Finnish Travel Agents, Ms. Heli Mäki-Fränti (personal conversation, 01.09.2017), for the travel and tourism/hospitality industry, the important megatrends until at least 2025 are globalisation, ageing population, technological development, and sustainable development. The description on sustainability as a megatrend is presented in Table 1.

Sustainability as a megatrend can be interpreted in different ways. It depends on the interpreter as well as well as their academic, professional, geographical and cultural background. Megatrends can be defined in diverse ways depending on how important it is to the receiver. Ageing population as an example of another global phenomenon, or megatrend is very relevant to Europe and part of Asia. But not that important to continents like Americas and Africa where the populations are growing. The previous statement is not totally valid, because the ageing population in other countries have impacts on export of products and services like tourism to the ageing markets . So, a megatrend can be seen or emphasised in diverse ways in different countries and regions. Megatrends can also have different impacts in different industries, and their importance can vary.

Sometimes industries emphasise a strong trend or an impact of a megatrend as a megatrend, such as a higher demand for health and wellness services in travelling. The demand of health and wellness services as such is not a megatrend, but is impacted by one (ageing population). Because it can be very important for the day-to-day business of a service supplier, it may be interpreted as a megatrend. Another megatrend, technological development, can be interpreted as increased connectivity or a significant rise in the demand for mobile services. Once again, the megatrend is technological development and its impacts are increased connectivity and demand for mobile services. Table 2 shows how sustainability as a megatrend is interpreted in this studying the framework of travel intermediation.

Outcomes of the expression of loyalty are intention of repurchasing, repurchasing, word of mouth (WoM), and brand appreciation. According to Beatty, Homer and Kahle (1988) a satisfied customer, does repurchasing and consequently forms loyalty. When the frequency of repurchasing intensifies, the customer is not searching much for alternative services any longer.

According to Schneider et al. (1998) service climate refers to customer service and service quality based on employees' perceptions of rewarded, supported and expected practices and behaviours. Service climate exits when a company's organizational climate focuses on supplying services to customers (Schneider et al., 1992). Deshpandé and Webster (1989) state that organisation culture represents the norms and values of an organisation, and the organisation climate describes how this culture is operationalised in the everyday tasks and activities. Both include the joint understanding of the organisation by the managers and employees. Company culture is interpreted by managers to employees, including those who work with the customers (Dean, 2004). It is possible to say that service culture extracts through all organisational levels of a

Table 2. Sustainability in travel intermediation

Mega Trend	Impacts on Travel and Service Industry	Impacts on Travel Intermediation Organisation	Changes in Organisations and Their Structures
Sustainability	Low-carbon and resource efficiency; enhancing the shift to more environmentally friendly technology and policies; mass tourism; over tourism	Virtual travelling, virtual meetings and technology. More environmentally friendly technology in aviation, shipping, and land traffic. Also, hospitality services accommodation & catering focus on sustainability. Trains replace flying in short and middle haul trips. Domestic and short distance travel more popular.	The generation gap and its importance in the work life: Generations Z & Y vs. generations X & baby boomers (Orange, 2016)

company from the top to the operational level where the customer meetings take place. Company culture also influences how customers see the company. Service climate is something collective, descriptive and service specific in its framework (Bowen & Schneider, 2014).

The antecedents of service climate are leadership, HRM practices, and systems support from operations, marketing, IT, and others. Leadership is central when constructing and maintaining a service climate. In leadership, the question is not always about vision and strategies, but also on everyday tasks. In the framework of service climate, relevant issues are, for example competent personnel with a positive attitude to customer service. System support related to service climate include equipment and machinery, and customer records, to name a few (Bowen & Schneider, 2014). He et al. (2011) suggest that service climate is a combination of three components: Customer orientation, managerial support and work facilitation. Customer orientation focuses on customer interests and its position as a high priority in business activities. Furthermore, it includes the generating and use of customer information for business purposes (Day, 1994.). Managerial support is defined as actions supporting and rewarding employees towards the delivery of quality service. Work facilitation is about general HR practices, work guidelines and supervision, as well as cooperation in the organisation to ease the delivery of quality service (He et al., 2011).

Another view on service climate is servant leadership, its impact on service climate and on customer service performance. Servant leadership describes a new organisational strategy that is enhanced by customers' concern about how to create a better world. Servant leadership puts the employees' need at the front and the concept of serving a broader community (Linuesa-Langreo & Elche-Hortelano, 2017). According to Trompenaars and Voerman (2009), the basic idea behind servant leadership is being human. They continue that the fast-changing world and work relationships must have impacts in the management and leadership models in organisations. In addition, Maitland and Thomson (2011) emphasise changes in work life that should be faced with new employment and management practices. They argue that change takes place in the new workforce, and that motivation are not always tied to money. The message is that the employer should capture the hearts, minds and wallets of the employees.

3. Data collection and survey results

The survey was executed in March and April 2018. It was sent to the member organisations of the Association of Finnish Travel Agents who forwarded it to their customers or people who were in their email databases. First, there were a few demographic questions concerning the gender, age, education and place of living of the respondents, followed by some questions about their travel habits. The demographic questions were multiple-choice questions and the questions about travel habits both in multiple-choice and open-ended formats. For the remaining questions respondents were asked to indicate their agreement with a number of statements, using a scale ranging from zero (total disagreement) to 10 (total agreement). The standard deviation (SD) showed a wide range for most of the statements. The distance between the mean/average and median displayed a wide range as well. Thus, the use of the median as a characteristic or key figure was more suitable than the mean/average (Heikkilä, 2014; Holopainen & Pulkkinen, 2002). During the analysis, only cases where the difference in the numerical value of the median is two or more are reported on, because it might have an importance when comparing the results and travel behaviour of the respondents.

The total number of respondents was 182 and 75 % of them were female. The two largest generation groups were theBabyBoomers II (1953–1965), also known as generation Jones (32 %)/g and Generation X, born in 1966–1980

Table 3. Generations and their birth years

Generation	Birth years
Silent generation	Before 1945
Baby Boomers I	1945 - 1952
Baby Boomers II, Generatikon Jones	1953 - 1965
Generation X	1966 - 1980
Generation Y	1981 - 1994
Generation Z	After 1994

(36 %). Most respondents lived with their spouses (41 %) or alone (29 %) and only about one out of five lived in families with children. The respondents were well-educated, with 70 % of the respondents having a higher-level diploma or degree (college or university). The respondents were working (62 %) or retired (27 %) and lived in towns and cities (85 %).

The respondents were rather experienced travellers while they have done on average at least three leisure trips during the previous two years. They were interested in new travel destinations, such as destinations in the Western Balkan, Africa and South America, and prepared to travel there as well. To travel to close-by or neighbouring destinations, e.g. domestic or Estonia, Russia and Sweden, in the near future did not raise much interest.

In general, it was not easy to find differences between individual groups by cross-tabulation, and the sample was too small for multi-variate analyses (Heikkilä, 2014). Thus, the following paragraphs aim to describe some differences, and in the end of this chapter personas to paint a deeper picture of the attitudes of Finnish travel agency customers. The personas were created by cross-tabulating the statements linked to sustainability and responsibility in the survey. There were eight statements, and the attitudes were divided into negative, neutral and positive. As mentioned above, the scale of the statements varied from zero to 10 (0 = totally disagree, 10 = totally agree), thus the values from zero to three were regarded negative, four to seven neutral and eight to 10 positive.

All the generations (Table 3) included in the analysis believed that a travel agency must conduct business in a sustainable way. They also found the well-being of the staff important when they chose an agency. This was appreciated most among the representatives of the baby boomers II. Travel agencies' values had a meaning when people booked a leisure trip. These values were the least important to the representatives of the Generation Y. All respondents wanted to purchase their trips in agencies who value their employees. Again, this was less important to the representatives of the Generation Y. Sustainable development was an important reason for all the respondents when they chose leisure travel services. This was least important to the Generation Y. When comparing the results by gender, the differences remained rather small. Females appreciated sustainability and responsibility of business when choosing the travel agency, a bit more than men. The well-being of the employees was especially important for people living in the capital area, but less important for respondents in the other big cities, mid-sized towns and municipalities. The values of the travel agency were emphasised among respondents of the small towns. Virtual travelling interested people living in the capital area the least, and people of the mid-sized and small towns the most.

Statement 14 in the survey was about respondents' interest in near-by destinations in the future: Mostly women were more interested in near-by destinations. They cover 84 % of the positives; men are more neutral. Baby Boomers I and Generation X were more interested in near-by destinations than the other generations, but Generation Y were also least interested in them. Furthermore, they were also the largest generation in the neutral scores. The most positive households were singles and respondents living with their spouses. The attitudes cross-tabulated by the education background show that the higher the education is the less interest there was to travel to near-by destinations. Respondents working were less interested in near-by destinations than retired respondents. There was no difference between respondents based on their place of living. People who have done three to five leisure trips in the last two years were most interested in near-by destinations when respondents who have done over 10 trips were least interested.

When quality is one element in sustainable and responsible business, statement 33 is part of this analysis. Females were less interested in doing business with an agency with good service, and males tend to be more neutral. Baby Boomers II appreciate good service and Generation X not that much. Good service was appreciated among families, singles were more neutral, and empty nesters (living with spouses) covered quite evenly all three groups. The higher the education the less appreciation there was for good quality service. Respondents with vocational and college degrees appreciated good quality service, respondents with a Bachelors degree were neutral, and those with a Masters degree were negative. Retired people did not appreciate agencies with good service to a large extent. People living in the capital area appreciated good service agencies less, while it is the other way around among respondents living in middle-sized towns. When it comes to the importance of the frequency of travelling, the least and most travelled did notpay much attention to the good services of a travel agency, but it was most appreciated among respondents who had done 6–10 leisure trips in the past two years.

Responsible business (statement 34) was important for women, but not for men. It was rather important for Baby Boomers II, neutral to Generation X, and least important to Generation Y. It was least important for singles, and to some extent important to respondents living with their spouses. People with a university degree were neutral, but those with a college diploma were negative. Responsibility was important for both people who were working and retired. Respondents living in the capital area were least interested in responsible business, people living in other locations were more positive and neutral. People who had done 1–2 trips in the past two years wereleast interested in the responsible business of an agency, others were either neutral or found it as positive.

The wellbeing (35) of the staff when choosing an agency was more important for women than men. It was more important to Baby Boomers I and II and to younger generations who were more neutral or negative. Singles and empty nesters seemed to appreciate the wellbeing of the staff slightly more than families. Again, the lower the education level was, the more the wellbeing of the staff was appreciated. Respondents with a Bachelor's degree were more neutral and those with a Master's degree were negative. Retired people agreed to the wellbeing of the staff better than working respondents. People living in municipalities did not see the wellbeing as very important, similar to those who lived in the capital area. Other respondent groups were more neutral. More travelled respondents did not always choose the travel agency according to the wellbeing of the staff. This was more important to respondents who had done 3–5 trips in the past two years.

The values of an agency (36) were more important to women than to men. It seemed to be important to Baby Boomers I, but Baby Boomers II are rather neutral; Generation X are more negative and Generation Y neutral. Agencies' values are important for empty nesters, but for the other groups it is more a neutral issue, and for families it was more negative. Values are least important to Masters as well as to people who work. They were most important to people living in the capital area and municipalities. For others, it was neutral. Frequency of travelling did not have an influence on the attitudes towards values.

The staff appreciation of agencies (37) was a neutral issue for both men and women. It was neutral to all the other generations except for Generation Y. It was most important to respondents living with their spouses, but rather neutral to the rest of the groups. The importance of staff appreciation was lowest among respondents with a Master's degree. It was more important to working people than retired respondents. Staff appreciation was also more important for people living in the capital area than for those in other locations, andleast important to people living in municipalities. It is least important for respondents who had made 3–5 leisure trips in the past two years, but more neutral in the other groups.

Sustainable development in tourism (38), when purchasing travel services was more important to women than to men. It was rather important to every other generation, except for Generation Y. Sustainable development was important to singles, but rather neutral among other households. There was no large difference between the educational backgrounds. People working did not see it as important as those who were retired. Respondents in the capital area thought that sustainable development in tourism was either an important issue or not important at all but in the other locations respondents saw this as a neutral issue. Sustainable development was least important for most frequent and unexperienced travellers and most important among respondents who had done 6–10 trips in the past two years.

Virtual travelling (45) was a rather neutral issue for both gender groups. It seemed to be more attractive to older than younger generations, and for singles. It was more neutral to people of every education group. Virtual travelling was more attractive to retired than people who work. It was least attractive to people living in the capital area than to those living in other cities and towns. Virtual travelling was most interesting for respondents who had done 6–10 trips in the past two years.

The attitudes of various customers towards sustainable tourism and responsible business is presented in Table 4. The average customer who thought positively about sustainability and responsible business in travel intermediation was a woman belonging to Baby Boomer generations, especially to Baby Boomers II. Baby Boomer I representatives were married and lived with her spouse and were retired. Baby Boomer II representative was single or lived with her spouse and was still working. She had a vocational or college diploma and lived in the capital area. She was a frequent traveller but did not belong to the most travelled group.

A customer with more a neutral attitude towards sustainability and responsibility in travel intermediation was a female representative of generation X who workedand lived in any location and had a Bachelor's degree. The household form also varied.

A customer li who did not find sustainability and responsible business that important was a working man with a university degree and who had travelled a lot in the past two years. He lived in the capital area. Another persona who

Table 4. Classification criteria for personas

Statement	Positive	Neutral	Negative
14 (nearby destinations)	Women; Baby Boomers I & II; singles, empty nesters; retired; 3–5 trips in the past two years	Generation X	Generation X; Bachelors & Masters; working; more than 10 trips in the past two years
33 (quality services)	Baby Boomers II; families, empty nesters; vocational & college education; middle-sized towns; 6–10 trips in the past two years	Men; singles, empty nesters; bachelors	Women; Generation X; empty nesters; Masters; retired; capital area; 1–2 & over 10 trips in the past two years
34 (responsible business)	Women, Baby Boomers II; empty nesters; working & retired;	Generation X; Bachelors & Masters	Men; Generation Y; singles; college diploma; capital area; 1–2 trips in the past two years
35 (staff wellbeing)	Women; Baby Boomers I & II; singles & empty nesters; vocational & college diploma; retired; capital area; 3–5 trips in the past two years	Bachelors	Men; families; Masters; working; municipalities; 6–10 & over 10 trips in the past two years
36 (agencies' values)	Women; Baby Boomers I; empty nesters; capital area	Baby Boomers II; Generation Y; singles; other locations	Men; Generation X; families; Masters; working
37 (staff appreciation)	Empty nesters; working & retired; capital area;	Men & women; other generations; other households; other locations	Generation Y; Masters; municipalities; 3–5 trips in the past two years
38 (sustainable services)	Women; other generations; singles; retired; capital area; 6–10 trips in the past two years	Other households; other locations; 3–5 trips in the past two years	Men; Generation Y; working; capital area; 1–2 trips & more than 10 trips in the past two years

(continued on next page)

Table 4. Continued

Statement	Positive	Neutral	Negative
45 (virtual travelling)	Baby Boomers I & II; retired; 6–10 trips in the past two years	Men & women	Generation X & Y; working; capital area

was less interested in sustainable development and responsibility was a woman belonging to Generation Y and who lived with her husband and children and had a university degree. She was also working and lived in the capital area or probably in a municipality. She had not travelled much in the past two years.

4. Discussion

The results of the survey showed that especially the question of sustainability was not of high importance to customers of Finnish travel agencies. It had importance when it concerned sustainability of travel services, like flights, accommodation and food services, However, Finns were not very interested in nearby destinations or virtual travel, both of which were in the focus of sustainable development in tourism. In this study, nearby destinations were domestic destinations and destinations in the neighbouring countries like Estonia, Norway, Russia and Sweden. Of course, Finns travel in Finland and to the neighbor countries, but these destinations are not going to become popular in the future. New destinations for Finns were especially countries in West Balkan, Africa and South America, thus the future demand will emphasise both European and long-haul holidays.

Responsibility was more in the hearts of the customers of the Finnish travel agencies. An appreciated agency had to have acceptable business values, it had to do business in responsible ways, and take care of the wellbeing of its employees. This might be, of course obvious, but the emphasis on the responsibility compared to sustainability is interesting.We can assume that responsible way to do business should increase customer loyalty. There were differences between individuals and groups based on the demography, even though the differences statistically seemed to be rather small.

According to megatrends, the younger generations should have been more sustainability and responsibility savvy than the older generations. For some reason, this survey did not reflect this. Baby Boomers I and II paid more attention to sustainability and responsibility than Generations X and Y. Especially, more neutral attitudes of Generation X and even negative attitudes of

Generation Y generated further questions, and suggested the need for further research with the a focus on these.

5. Conclusion

This study has provided a profile of Finnish travel agencies and tour operators in terms of attitudes of their customers in sustainability and responsibility in the travel intermediation business. However, the number of statements was not very numerous. In the end, it was possible to get a picture of contemporary customers and their attitudes to exploit it to some extent in the business development. In the travel intermediation business in Finland, development focuses should be in sustainability especially in travel services and appreciation of stakeholders throughout the service chain.

As the sample was rather small, and the differentiation of various consumer groups challenging, the question of the reliability and validity was obvious. The population was reached, but according to the statistics given by the online tool, Webropol, only about 15 % of those who opened the survey replied to it. Those who started to reply finished it as well. There were many reasons for not replying, but according to the Association of Finnish Travel Agents it was a trend. The association gets less and less replies to their own surveys as well. In the future, the members of the association should be more active in the submission of the surveys of the association, and make answering to them more attractive.

References

Alamdari, F. (2002). Regional development in airlines and travel agents relationship. *Journal of Air Transport Management, 8*(5), 339–348.

Beatty, S. E., Homer, P. & Kahle, L. R. (1988). The involvement-commitment model: Theory and implications. *Journal of Business Research, 16*(2), 169–183.

Bowen, D.E. & Schneider, B. (2014). A service climate synthesis and future research agenda. *Journal of Service Research, 17*(1), 5–22.

Castillo-Manzano, J. I., & López-Valpuesta, L. (2010). The decline of the traditional travel agent model. *Transportation Research Part E: Logistics and Transportation Review, 46*(5), 639–649.

Cheyne, J., Downes, M., & Legg, S. (2005). Travel agent vs. internet: What influences travel consumer choices? *Journal of Vacation Marketing, 12*(1), 41–57.

Day, G. S. (1994). The capabilities of market-driven organizations. *Journal of Marketing*, Vol. 58, No. 4, 37.

Dean, A. M. (2004). Links between organizational and customer variables in service delivery: evidence, contradictions and challenges. *International Journal of Service Industry Management*, Vol. 15, No. 4, pp. 332–350.

Deshpandé, R. & Webster, F. E. Jr. (1989). Organisational culture and marketing: defining the research agenda. *Journal of Marketing*, Vol. 53, No. 1, 3–15.

Dilts, J., & Prough, G. (2002). Travel agencies: A service industry in transition in the networked economy. *The Marketing Management Journal*, 13(2), 96–106.

Dwyer, L., Edwards, D., Mistilis, N., Roman, C., & Scott, N. (2009). Destination and enterprise management for a tourism future. *Tourism Management*, 30 (2009), 63–74.

He, Y., Li, W., & Lai K.K. (2011). Service climate, employee commitment and customer satisfaction. *International Journal of Contemporary Hospitality Management*, Vol. 23, No. 5, 592–607.

Heikkilä, T. (2014). *Tilastollinen tutkimus.*Edita.

Holopainen, M., & Pulkkinen, P. (2002). *Tilastolliset menetelmät.*WSOY.

Horwath HTL (2016). *Tourism Megatrends. 10 things you need to know about the future of Tourism.* Retrieved at www.horwathhtl.com on 13 September 2017.

Linuesa-Langreo, J., & Elche-Hortelano, D. (2017, November 15). *New Strategies in the New Millennium: Servant Leadership as Enhancer of Service Climate and Customer Service Performance.* Frontiers in Psychology. https://www.frontiersin.org/articles/10.3389/fpsyg.2017.00786/full

Lominé, L., & Edmunds, J. (2007). *Key concepts in tourism.*Palgrave.

Lubbe, B. (2000). *Tourism distribution: Managing the travel intermediary.* Jutta and Co.

Maitland, A., & Thomson, P. (2011). *Future Work. How Businesses Can Adapt and Thrive in the New World of Work.* Palgrave MacMillan.

Novak, J., & Schwabe, G. (2009). Designing for reintermediation in the brick-and-mortar world: Towards the travel agency of the future. *Electronic Markets*, 19, 15–29.

OECD (2017). *ISSUES PAPER: Analysing Megatrends to Better Shape the Future of Tourism.* OECD High Level Meeting on Tourism Policies for Sustainable and Inclusive Growth.

Orange (2018, July 31). *What role for HR in 2020–2025?* http://www.oliverwyman.com/content/dam/oliver-wyman/global/en/2016/june/What%20role%20for%20HR%20in%202020-2025.pdf

Oxford Economics (2018, July 31). *Shaping the Future of Travel*. https://amadeus.com/documents/en/airlines/research-report/oxford-economics-shaping-the-future-of-travel.pdf

PWC (2016, September 20). *Five Megatrends and Their Implications for Global Defence & Security*. www.pwc.uk/megatrends

Ritalahti, J. (2018). From High Street to Digital Environments: Changing Landscapes in Travel Intermediation. In A. Ali & J. Hull (Eds.), *Multi-stakeholder perspectives of the tourism experience: Responses from the International Competence Network of Tourism Research and Education (ICNT)* (pp. 135–147). Peter Lang Verlag.

Scaglione, M., & Schegg, R. (2016). Forecasting the final penetration rate of online Travel Agencies in different hotel segments. In R. Schegg & B. Stangl (Eds.), *Information and Communication Technologies in Tourism 2016* (pp. 709–721). Springer Verlag.

Schneider, B., Wheeler, J. K., & Cox, J. F. (1992). A passion for service: using content analysis to explicate service climate themes. *Journal of Applied Psychology*, Vol. 77, No. 5, 705–716.

Schneider, B., White, S. S., & Paul, M. C. (1998). Linking service climate and customer perceptions of service quality: Test of a causal model. *Journal of Applied Psychology, 83*(2), 150–163.

Skift (2018, January 20). *Skift Megatrends of 2017*. https://skift.com/2017/01/12/the-megatrends-defining-travel-in-2017/

Trompenaars, F., & Voerman, E. (2009). *Servant-leadership across cultures*. Infinite Ideas Limited.

Anne Köchling, Julian Reif and Rebekka Weis

Tourism and Political Crises: An Analysis of British Holiday Planning in Times of "Brexit"

1. Introduction

On 23 June, 2016, a referendum was held in the United Kingdom (UK) in which the citizens were given the chance to vote on whether the UK would remain a member of the European Union (EU). Fifty-two percent of the British voted for the UK to leave the EU; the voter participation, at 72 %, turned out to be quite high in comparison with the last parliamentary elections (Elmer et al., 2016). The background for the referendum was the increasing dissatisfaction of the British with the "open borders" policy of the EU (Theurer, 2016a) as well as dissatisfaction with the push for stronger interdependence between the EU member countries (dpa, 2016). The financial markets had already noticed a sizable negative impact on the value of the British pound a few days before the EU referendum (Theurer, 2016b). This trend continued after the decision was made. In the autumn of 2016, the lowest value of the pound since 1985 was recorded in the wake of a speech by the new Prime Minister, Theresa May, in which she announced her planned hard course for the EU exit negotiations (Zeit online, 2016).

In addition to the fundamental political and economic consequences, the impact of the EU referendum on the tourism market has been controversially debated as British guests form a major target group for many EU countries. The fallen exchange rate for the pound against the Euro, in particular, makes trips to other European countries less attractive for the British. Thus, a decline in British tourist numbers is expected as a local effect on tourism, especially in the southern European countries, which is not to be underestimated (Lang, 2016). Even in a survey that was answered by decision-makers in the German travel industry in July 2016, 72 % of the respondents agreed or mostly agreed with the statement that travel to the continent would become too expensive for the British. At the same time, the majority of the experts (74 %) agreed or mostly agreed that travel to the UK will become significantly cheaper, and thus they expect a rise in demand to occur (Travel Industry Club & manufacts research & dialog, 2016).[1] With regard

---

[1] The survey was a non-representative online survey carried out from 6 - 13 June 2016 (n=147) and was commissioned by the Travel Industry Club, an independent business club, founded in 2005, as well as by representatives of the travel industry (see Travel Industry Club & manufacts research & dialog, 2016).

to freedom of movement, on the other hand, no changes will take effect until the UK has left the EU (Foreign & Commonwealth Office, 2017), so the effects which could justifiably be attributed to that will not be felt until 2019 at the earliest.

Against this background, the Institute for Management and Tourism (IMT) of the West Coast University of Applied Sciences examined the British holiday planning for 2017 in cooperation with the German National Tourist Board (Deutsche Zentrale für Tourismus e.V. (DZT)) in January 2017. The aim was to examine whether the upcoming Brexit would have an impact on British travel behaviour in 2017 in general and holiday travel behaviour to Germany in particular.

2. Crises and their Impact on Tourism Demand

In addition to external shocks to the tourism industry caused by terrorist attacks and acts of violence, current political and economic crises also play a role in the discourse on the impact on tourism demand. For example, declines in travel bookings to the United States were observed after the election victory by US President Donald Trump and the short-term ban on the entry of citizens from seven countries around the world, which were mainly characterised by Islam (Forward Data, SL, 2017).

In the following section, the term Brexit is conceptualised as a political crisis and the effects of crises on tourism demand are discussed. By definition, crises affect the normal functioning of tourism companies due to the negative impression of the affected destination that has an impact on the potential travellers:

> (...) [tourism crises] threaten the normal operation and conduct of tourism-related business; damage a tourist destination's overall reputation for safety, attractiveness, and comfort by negatively affecting visitors' perceptions of that destination; and, in turn, cause a downturn in the reduction in tourist arrivals and expenditures." (Sönmez et al.,1999, p. 14)

The negative perceptions, which are quickly disseminated through the media, can be impressively demonstrated by the following figures: In the year 2012 and thus at the time of the Greek financial crisis, more than a quarter of the Germans (26 %) stated aspects referring to "Economy, Politics, Crisis" in a survey which asked for spontaneous associations with Greece as a travel destination (IMT, 2012).[2] The specific topics "Crisis / Euro Crisis / Debt Crisis / Economic

---

2   Survey period: October 2012; n=1,000, representative of the German-speaking population aged 16 – 70 years (see IMT, 2012).

Crisis" were mentioned by almost 7 % of the respondents in association with Greece as a travel destination. Although the negative reporting on the financial crisis in Greece led to a clearly measureable result in the spontaneous associations with the travel destination among the Germans, the natural strengths of the holiday country were still dominant (e.g. 17 % of the respondents associated Greece with "sun / sunny", 13 % with "(beautiful, blue) sea") (IMT, 2012).

In a more general sense, crises can be categorised as being either endogenous or exogenous (Schroeter, 1996; Dreyer et al., 2001; Petermann et al., 2006): Whereas endogenous crises are human induced (e.g. strike) or have technical causes (e.g. technical failure in aviation), exogenous crises have an external effect on the tourism stakeholders – in the source region, during the journey and in the destination zone. In addition to geophysical factors (e.g. volcanic eruption), socio-cultural factors (e.g. demonstrations), religious factors (e.g. terrorist attacks with a religious background), medical factors (e.g. epidemics), and politico-economic influencing factors (e.g. trade conflicts) could all play a role in exogenous crises (Schroeter, 1996).

From the perspective of the EU countries, which is the focus of the present study, Brexit can be seen as an exogenous, politico-economic crisis in which the EU countries, and thus also the Federal Republic of Germany, could suffer image losses in the UK. This could result in a reduction in the intensity of foreign travel by the British to these countries, with a direct effect on the tourism-induced value, among others. At the same time, it can be argued that the image of the UK as a destination for Europeans could also change as a result of leaving the EU. However, this view of the impact on inbound tourism in the UK is not the focus of this article. Nonetheless, 9 % of the German-speaking population aged between 14 – 70 years old claimed that Britain's vote on the Brexit influenced their holiday travel in 2016 (Lohmann, 2017). What impact this had on e.g. cancellations, or the like, remains unanswered. Depending on the perspective, this crisis could also be viewed as an opportunity. In addition to the existing potential to create innovations from crises (Dreyer et al., 2001), it can be assumed that the British tourism industry is favourably inclined towards Brexit as it could profit from the expected increased demand in the UK.

The impact of crises on the choice of destination and travel behaviour of the consumers is difficult to predict due to the subjectively different risk perceptions. Divergent risk perception and, ultimately, a realised or not realised travel-behaviour, is due to a complex interaction of the systems of humanity, media, and tourism (Dreyer et al. 2001). The two basic determinants of the travel decision, travel-willingness (e.g. desire to travel) and the ability to travel (e.g. money, health, political conditions in the destination and source regions) (Lohmann, 2016), meet the (perceived) media stimuli, for example the advertising of destinations, and the individual personal factors of the travel decision: Cognitive determinants (e.g. perception of reports in the mass media), activating determinants (e.g. travel needs and motives) as well as personal determinants (e.g. risk affinity) (Dreyer, 2000).

The Brexit crisis has a direct impact on travel ability and can be viewed as a tourist crisis of demand. The case is similar to the financial crisis in Greece, where the Greek were essentially lacking the money to travel in their own country (Lohmann, 2016). The difference here is that with the impending withdrawal from the EU, it can be assumed that the British will be inclined to travel *more frequently* in their own country due to the depreciation of the pound.

In contrast to the influence of politico-economic crises, the impact of terrorist attacks on the corresponding target region and the demand has been well-researched and documented (e.g. Sönmez, 1998; Kuschel & Schröder, 2002; Aschauer, 2009; Lohmann, 2016). Terror attacks are spatially and temporally limited (Lohmann, 2016) and, due to global multi-selectivity in destination choice, lead to high volatility in tourism demand (acting substitution effects) (Steiner et al., 2006). Although the overall global demand during crises – with the exception of the global financial crises starting in 2007 – remains relatively stable with regard to the affected destination, the travel flows can shift significantly.

In the case of politico-economic or social crises, which in comparison have a rather creeping development from a temporal point of view (Dreyer et al., 2001), a further parallel effect can be identified. Regionalisation and intensification processes can lead to the development of domestic tourism in their own country in the wake of great uncertainty on the part of the consumer. Investigations after the 9/11 terror attacks in the USA show that regionalisation processes had an effect in the Western world on the one hand (due to increased risk awareness with regard to Islam among others) and in the Arab world on the other hand (inter alia because of the fear of having to expose themselves to hostility abroad) (Steiner et al., 2006). In the tourism year 2015/2016 there was also an increase in domestic tourism in Germany. Compared to the previous year, the growth in private domestic overnight stays can be explained by (1) intensification effects (the Germans were travelling more in their own country and were also staying longer) and (2) Germany profiting from losses of foreign destinations (Germans, who had spent a private night abroad in the previous year, were travelling in Germany during that period) (GfK & IMT, 2017a). The extent to which this growth is related to the uncertain political situation in destinations abroad, such as Turkey, during the investigation period is not proven, but can however be assumed. With regard to Brexit, the question arises as to whether such regionalisation processes are also to be observed and whether the British are increasingly travelling within their own country.

3. The Travel Behaviour of the British

In 2016, the British undertook 67.8 million trips abroad with at least one overnight stay (DZT / IPK, 2017). For these overnight stays abroad, Spain, France and the United States were the most popular destinations for the British in 2016. Germany was in the sixth place, with 3.1 million visits by the British and a market share of 5 %. The majority of the trips to Germany (60 %) were holiday trips, 24 % were business trips and 16 % were visits to relatives/friends and other trips. Revenues generated by trips of the British to Germany were at 2.5 billion Euro in 2016 which have risen by 47 % since 2012 (1.7 billion Euro) (DZT / IPK, 2013–2017).

According to the figures collected by the Federal Statistical Office concerning overnight accommodation volumes in commercial establishments with 10 beds or more, the United Kingdom is the fourth most important inbound market for Germany, after the Netherlands, Switzerland and the USA, with a volume of 5.6 million overnight stays in 2016. The annual number of overnight stays has grown by 23 % since 2012 (4.5 million) (Statistisches Bundesamt, 2013–2017). Figure 1 provides an impression of the monthly development of this source market since 2010.

When looking at the monthly time lapse of the officially recorded overnight stays in commercial establishments with 10 beds or more in Germany from the United Kingdom since January 2010, the negative change rates are particularly noticeable in the period from May to October 2016. Although there were temporary negative rates of change in the summer of 2011 and in May 2013 and at the beginning of 2014, the declines in May 2016 (-3.3 %) and especially in August (-8.1 %) were the highest in six years. Despite a renewed rise in overnight stays in November and December 2016, the question arises as to whether these declines can be explained by the Brexit vote or indirectly to the related fall in the exchange rate of the British pound against the Euro.

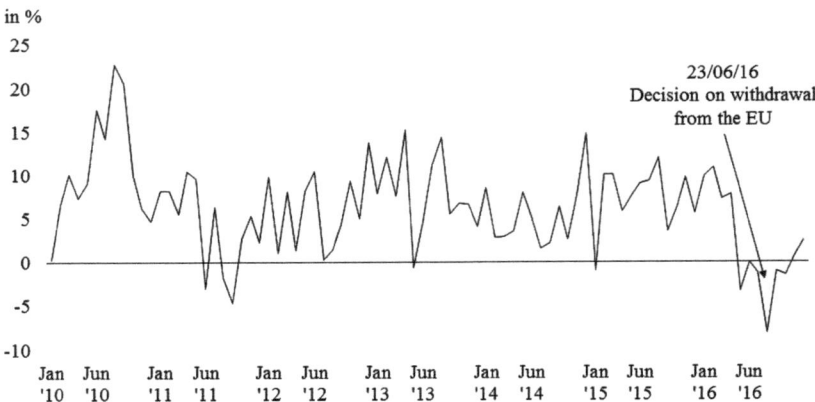

Fig. 1: Rate of change of the officially recorded overnight stays in Germany from the United Kingdom. Source: Calculation based on figures from the Federal Statistical Office (Statistisches Bundesamt) 2017. Rate of change (in %) of officially recorded overnight stays in commercial accommodation establishments with 10 beds or more from the source market of the United Kingdom compared with the same month in the previous year (January 2010 – December 2016).

Finally, regarding some qualitative aspects, the British are rather satisfied with their holidays in Germany. According to the latest guest survey British holidaymakers rated the overall satisfaction with their stay in Germany from May 2015 until October 2016 at 1.7 points on a 6-point-scale from "very good (1)" to "unsatisfactory (6)". Accommodation and gastronomy were rated at 1.9 points, and the price-performance-ratio at 2.0 points. Compared to the evaluation of all holiday guests in Germany from abroad the satisfaction was slightly above average (e.g. overall satisfaction: 1.8 points) (DZT, 2015/2016).

4. Research Question and Methodology

The focus of this study was to examine whether Brexit had an impact on the holiday travel plans of the British in 2017 in general and on the choice of travel destinations in particular. In contrast to the above-mentioned officially recorded overnight stays from the source market of the United Kingdom, which included both private and business trips, the study at hand focussed on all planned holiday trips by the British with at least one overnight stay in 2017.

An online panel survey was chosen as the survey method due to financial restraints. The survey was carried out by the Ipsos Field Institute and based on their Access Panel in the UK. Therefore, the base population consisted of English speaking internet users between 16 to 75 years of age that were members of the Ipsos Access Panel in the UK. A quota sample of 2,000 people representative of the population of the UK with regard to age, gender and region was drawn from the panel. The relatively high sample size was selected in order to have sufficient cases for in-depth analysis e.g. of sociodemographic differences.

The survey took place from 13 to 17 January 2017. A speech by Theresa May concerning Brexit was also expected for the last day of the survey so that the topic was very present in the media during the survey period. At the same time, the exchange rate of the British pound against the Euro fell once again to the level of the October trough (Spiegel online, 2017).

For data analysis the program IBM SPSS Statistics was used. In addition to the univariate frequency counts for the descriptive data analysis, the statistical significance of all cross tabulation results was tested using chi-square-test at a significance level of 0.05. All results presented in the following chapter show a p-value of $p < 0.05$.

5. Impact of Brexit on British Holiday Planning 2017

5.1 *British Holiday Planning 2017*

Nearly two thirds of the British (65 %) planned holiday trips with at least one overnight stay in 2017. Especially eager to travel were women (70 %), the age group between 55 and 75 years old (71 %), respondents with a high level of education, i.e. NVQ5[3] or degree (78 %), living in 2-person households (72 %) and those with an annual household net income of £25,000 or more (approx. €29,000). Every fourth respondent (26 %) did not plan on travelling on holidays in 2017. In addition, there was a small group (9 %) of undecided respondents (Response: "Do not know").

When asked whether more, the same or fewer trips with at least one overnight stay was planned in 2017 in comparison with the previous year, 63 % of the British planning to travel answered with "the same amount". One in four of the respondents (25 %) planned to go on even more holiday trips than in 2016

---

3   NVQ: National Vocational Qualifications; Professional qualification in Great Britain on five levels. Level 5 means that there is the minimum of a Master's qualification (highest level of education).

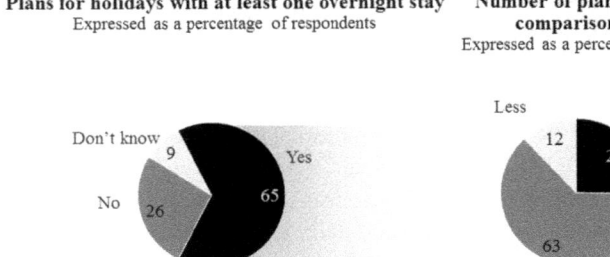

Fig. 2: British holiday planning 2017. Source: Authors

(see Figure 2). Among those respondents, the younger age group, from 16 to 24 years of age, was overrepresented (36 %). Only 12 % of the respondents said they wanted to travel less in 2017 than in the previous year.

The undecided and those who do not plan a holiday trip are particularly interesting for the estimation of the Brexit effect. For this reason, the group of undecided respondents were asked whether the uncertainty about their holiday planning was related to the upcoming United Kingdom exit from the EU. Only 14 % of the undecided answered this question with "Yes".

The sub-group of those that do not plan a holiday trip in 2017 were asked for their reasons. More than half (54 %) of the respondents said that they could not afford a holiday trip financially. In the North West (70 %) as well as the South West and Wales (61 %) regions, the proportion of respondents with this justification was significantly higher. However, a direct link to the weak exchange rate of the pound against the Euro could not be found in the data. For example, the proportion of those with an annual household net income below £20,000 was clearly overrepresented at 47 % compared to the total population (24 %), which means that the disposable income could be used as an explanatory variable. With a large gap followed the reasons "I don't travel in general" (16 %) and "I generally prefer to stay at home because that is where I can relax most" (15 %). The global threat of terrorism (7 %) as well as Brexit (4 %) played only subordinate roles as reasons for "Not travelling" (see Figure 3).

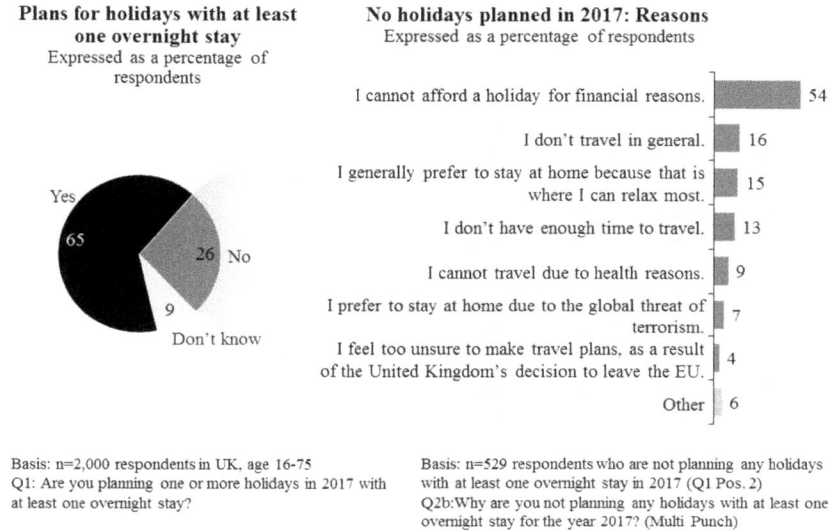

Fig. 3: Reasons not to travel on holidays in 2017. Source: Authors

5.2 British Holiday Destinations 2017

Nearly half of the British that were willing to travel (49 %) were planning a holiday trip with at least one overnight stay in their own country for 2017. Especially the older age group between 55 and 75 years of age (57 %), those in 2-person households (54 %) and the inhabitants of the Midlands (56 %) intended to go on holiday in their home country.

Of the British who planned to spend their holidays in the United Kingdom, almost half intended on solely staying in the United Kingdom in 2017. The main reasons were that there was still so much to discover in the country (45 %) and the domestic tourism industry should be supported (26 %). Those who were older than 45 years of age especially agreed with these reasons. A fifth (21 %) of the "Only UK holidays" respondents cited the global threat of terrorism as a reason to stay in their home country. The impending exit from the EU and fears that a trip to the EU would become too expensive (16 %) or a general feeling of insecurity because of this (9 %) were mentioned as reasons by only a small proportion of respondents (see Figure 4).

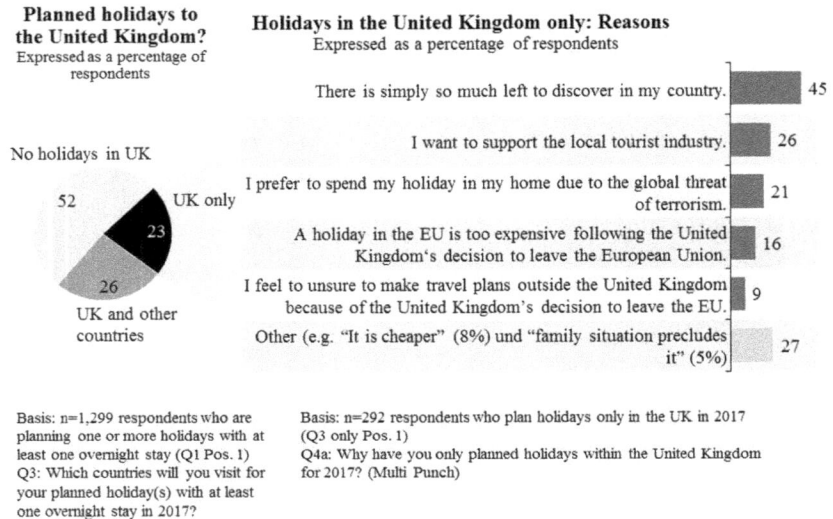

Fig. 4: Reasons for solely taking domestic holidays in 2017. Source: Authors

The foreign destinations for current holiday planning showed a similar order of popularity as the foreign travel destinations of the past:[4] Spain (22 %), France (13 %) and the USA (11 %) formed the three most popular foreign travel destinations for British holiday travellers in 2017. At least one holiday trip to Germany was planned by 5 % of the British who were willing to travel in 2017 (see Figure 5).

5.3 *Holiday Destination Germany*

In addition to the general holiday planning of the British, the study also focused on the target market for Germany. Those who had holiday plans for Germany (5 % of the British who were willing to travel) were asked about the number of planned holiday trips to Germany. The overwhelming majority (82 %) of those who wanted to go on a holiday trip to Germany in 2017 were only planning a single trip to Germany.

---

4 See the main destinations of the British when travelling abroad 2016: DZT / IPK 2017.

Destinations 2017
Expressed as a percentage of respondents

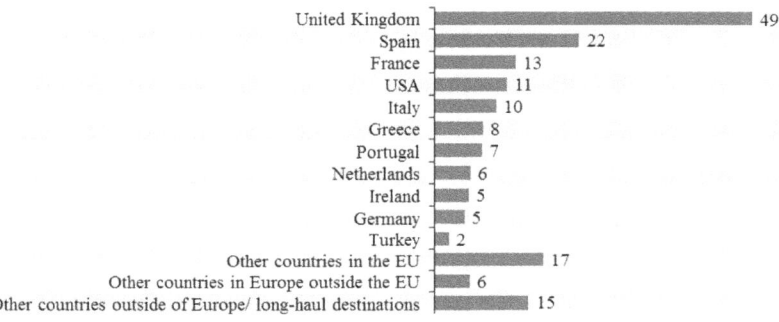

Basis: n=1,299 respondents who are planning one or more holidays with at least one overnight stay (Q1 Pos. 1)
Q3: Which countries will you visit for your planned holiday(s) with at least one overnight stay in 2017? (Multi Punch)

Fig. 5: British holiday destinations 2017. Source: Authors

Furthermore, the British who were planning at least one holiday trip in 2017, but who were not planning a journey to Germany, were asked why they did not want to travel to Germany. About one in three (35 %) said that they had not yet considered a trip to Germany. This reason was particularly frequently given by the young age group up to 34 years old (41 %). The second most common reason was that other destinations were more attractive (28 %), followed by the statement that Germany was not an interesting destination (22 %). Older British travellers between 55 and 75 years of age (26 %), unemployed respondents (27 %) as well as the inhabitants of the North and Yorkshire regions (30 %) were the groups that especially considered Germany to be an uninteresting destination. The lowest amount of agreement with this statement was in Scotland (14 %) and Greater London (16 %). All other reasons received less agreement – Brexit was given as a reason for not wanting to travel to Germany by only 2 % of the respondents (see Figure 6).

In addition, the groups of "non-travellers" and the "uncertain" respondents were asked to comment on Germany as a holiday destination. Firstly, this group was asked whether, in general, Germany could be a holiday destination for them. Almost half (49 %) of the group answered this question with "Yes". The number of respondents who selected Germany as a potential holiday destination increased significantly with increasing educational levels (NVQ5 or degree: 71 %).

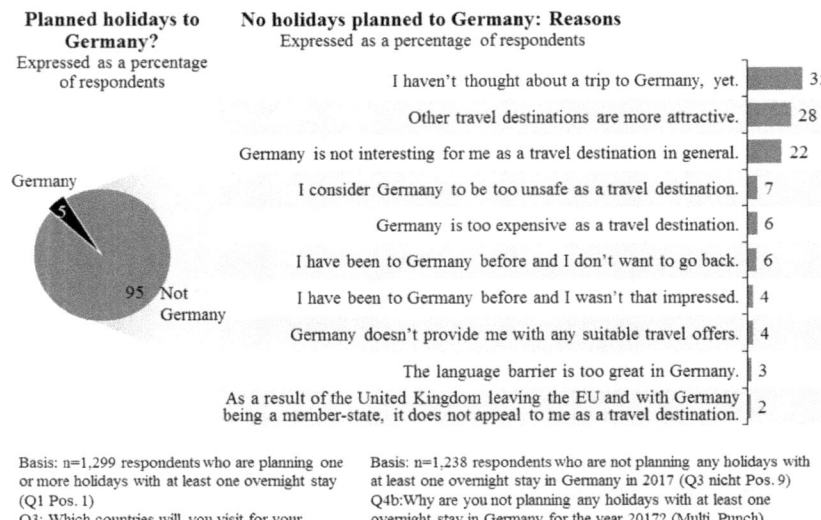

Fig. 6: Reasons for not spending the holidays in Germany in 2017. Source: Authors

Finally, those who would never consider Germany as a holiday destination were asked for their reasons for this. The most frequent answer here was that other destinations were more attractive (37 %). Moreover, even among the "non-travellers" and the "uncertain travellers", every third respondent (32 %) said that they had not yet considered a trip to Germany. There was a large gap to the third most frequently named reason which was that Germany is too expensive as a holiday destination (9 %). Brexit also played a very minor role here (5 %) (see Figure 7).

Germany is not considered as a travel destination in general: Reasons
Expressed as a percentage of respondents

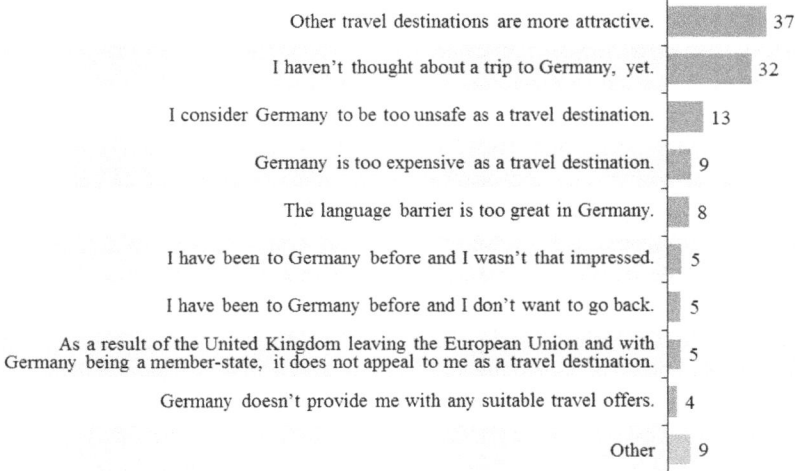

Reason	%
Other travel destinations are more attractive.	37
I haven't thought about a trip to Germany, yet.	32
I consider Germany to be too unsafe as a travel destination.	13
Germany is too expensive as a travel destination.	9
The language barrier is too great in Germany.	8
I have been to Germany before and I wasn't that impressed.	5
I have been to Germany before and I don't want to go back.	5
As a result of the United Kingdom leaving the European Union and with Germany being a member-state, it does not appeal to me as a travel destination.	5
Germany doesn't provide me with any suitable travel offers.	4
Other	9

Basis: n=367 respondents who would not consider Germany as a holiday destination in general (Q5 Pos. 2)
Q6: Why wouldn't you consider Germany as a holiday destination in general? (Multi Punch)

Fig. 7: Reasons why Germany would generally not be considered as a holiday destination. Source: Authors

6. Conclusion

Overall, the survey shows that the holiday mood of the British was quite positive at the beginning of 2017 and only a small proportion of respondents were planning fewer holiday trips than in the previous year. Only a few people said that Brexit was the reason for this uncertainty in their holiday planning or for the fact that they were not planning any holiday trips. Fundamentally, financial aspects played a significant role in British travel plans, as financial reasons were named by more than half of the respondents who were not planning a holiday.

Regionalisation effects, following the Brexit resolution, cannot be observed at present. Although about 50 % of British travellers were planning to go on holiday in their own country,[5] historically domestic tourism has always been

---

[5] The possible comparative value for Germany is 39 % taking into account the travel plans of the German-speaking online population aged 14 years old and older (49.2 mil. people) who are planning a holiday trip during the period between Nov. 2016 – Oct. 2017 with at least one overnight stay within Germany (see GfK / IMT, 2017b).

very important in Great Britain.[6] Furthermore, other studies based on travel planning predicted an increase in the number of British travelling to foreign countries for the year 2017 (IPK 2017, p. 10).

Brexit seems to have little influence on the holiday travel behaviour of the British travelling to Germany at the beginning of 2017. The recent positive changes in the number of officially recorded overnight stays as well as the ongoing holiday travel plans to Germany turn the Brexit crisis, with regard to the intensity of its effect, at least until the full exit from the EU, into a "V crisis" with a punctual, short-term slump in demand followed by a subsequent recovery.[7] For Germany, however, there is a general challenge to be anchored even more strongly as a potential holiday destination in the minds of the British. At the moment, from a British point of view, the travel destination Germany is less attractive in comparison to other destinations: just over a quarter of the British perceived Germany as a likeable tourist destination – a value that is clearly below average, even though the opinion is reciprocated (25 % of the Germans view Great Britain as a likeable travel destination) (Reif & Eisenstein, 2017). Nevertheless, once the British do visit Germany they are very satisfied with their holidays (DZT, 2015/2016).

Finally, it can be said that the upcoming Brexit does not (yet) keep the British from going on holidays. For the countries in which the British are an important target group in the holiday market there is, therefore, no short-term reason to worry. Nevertheless, the question arises of how the United Kingdom will adjust economically following a successful Brexit. A prolonged economic crisis, as well as possible restrictions on the freedom to travel, would certainly have a greater impact on the holiday travel market. Therefore, a repetition of the study is planned following a successful Brexit.

References

Aschauer, W. (2009). Krisen im Tourismus als neue Forschungsrichtung. Forschungsstand zu den Ursachen und Effekten terroristischer Anschläge. *Zeitschrift für Tourismuswissenschaft, 1(1)*, 13–28.

---

6 The percentage of personal/private travel by the British in their own country in 2012 and 2013 was 67 % and 66 % respectively (See Eurostat,2017a and Eurostat, 2017b.
7 Possible classification of crises on the basis of their temporal progression and the intensity of the effect in the form of a V, U shape (a slow slump in demand with slow consolidation), L shape (a persistent slump in demand with consolidation at a low level) and a W shape (successive slumps in demand; consolidation is interrupted) (see Schuckert & Möller, 2005 and Krystek, 1987).

Deutsche Zentrale für Tourismus e.V. (DZT) / IPK International (IPK) (2013–2017). *World Travel Monitor 2012–2016.* Excerpt of the results provided by DZT via e-mail on 27.04.2017.

Deutsche Zentrale für Tourismus e.V. (DZT) (2015/2016). *Qualitätsmonitor Deutschland-Tourismus.* Excerpt of the results provided by DZT via e-mail on 27.04.2017 and 05.03.2017.

Dreyer, A. (Ed.) (2000). *Kulturtourismus.* Munich, Germany: Oldenbourg.

Dreyer, A., Dreyer, D. und Obieglo, D. (2001). *Krisenmanagement im Tourismus. Grundlagen, Vorbeugung und kommunikative Bewältigung.* Munich, Germany and Vienna, Austria: Oldenbourg.

Elmar C. et al. (2016, June 24). Britisches EU-Referendum: Hier haben die Brexiteers gepunktet. *Spiegel online.* Retrieved from: http://www.spiegel.de/politik/ausland/brexit-ergebnisse-in-der-analyse-hier-haben-die-befuerworter-gepunktet-a-1099514.html.

Foreign & Commonwealth Office (Ed.) (2017, February 6). *Advice for British nationals travelling and living in Europe, following the result of the EU referendum.* Retrieved from: https://www.gov.uk/guidance/advice-for-british-nationals-travelling-and-living-in-europe.

Forward Data SL (Ed.) (2017, February 17). *Trump Travel Ban Impact on Air Travels to the U.S.A.*.Retrieved from: https://forwardkeys.com/revenue-management/article/trump-travel-ban-impact-on-air-travels-to-the-USA.html.

GfK SE, Shopper (GfK) und Institut für Management und Tourismus (IMT) (2017a). *Wanderungsanalyse – Inlands- und Auslandsreisen aus dem deutschen Quellmarkt.* [unpublished report]. Nürnberg, Heide, Germany.

GfK SE, Shopper (GfK) und Institut für Management und Tourismus (IMT) (2017b). *GfK / IMT DestinationMonitor Deutschland, Reiseplanungen der Deutschen.* Nürnberg, Heide, Germany.

Institut für Management und Tourismus (IMT) (2012). *Spontanassoziationen 2012 zu Griechenland als Reiseziel.* [unpublished report]. Heide, Germany.

IPK International (IPK) (2017). *ITB World Travel Trends Report 2016 / 2017.* Retrieved from: http://www.itb-berlin.de/media/itb/itb_dl_all/itb_presse_all/World_Travel_Trends_Report_2016_2017.pdf.

Kuschel, R. and Schröder, A. (2002). *Tourismus und Terrorismus. Interaktionen, Auswirkungen und Handlungsstrategien.* Dresden, Germany.

Krystek, U. (1987). *Unternehmungskrisen – Beschreibung, Vermeidung und Bewältigung überlebenskritischer Prozesse in Unternehmungen.* Wiesbaden, Germany.

Lang, F. P. (2016). Der »Brexit«: Rückschritt für Europa in der Welt, den die EU verkraften kann. *Ifo schnelldienst, 10/2016*, 21–24.

Lohmann, M. (2016). Terror und touristische Nachfragetrends. In FUR Forschungsgemeinschaft Urlaub und Reisen e.V. (Ed.), *Urlaubsreisetrends 2025. Entwicklung der touristischen Nachfrage im Quellmarkt Deutschland. Update 2016* (pp. 24–35). Kiel, Germany.

Lohmann, M. (2017, March 8). *Reiseanalyse 2017 – Präsentation der Ersten Ergebnisse, Pressekonferenz der Forschungsgemeinschaft Urlaub und Reisen e.V. (FUR)*. Retrieved from presentation on the ITB in Berlin.

Petermann, Th., Revermann, Ch. und Scherz, C. (2006). *Zukunftstrends im Tourismus, Büro für Technikfolgen-Abschätzung beim Deutschen Bundestag (TAB)*. Berlin, Germany.

Reif, J. and Eisenstein, B. (2017). Deutschlands Image als Reiseziel. Von Gastfreundschaft, Bergen und Bier. In Eisenstein, B., Schmudde, R., Reif, J. and C. Eilzer (Eds.), *Tourismusatlas Deutschland (pp. 18–19)*. Konstanz, Germany: UVK.

Schroeter, G. (1996). *Krisen-Management: Ein Leitbild zu einer kompatiblen betrieblichen und öffentlichen Gefahrenabwehr*. Ingelheim, Germany: SecuMedia.

Schuckert, M. R. and Möller, C. (2005). Krisenantizipation und -reaktion in der Touristik am Beispiel von Luftverkehrsunternehmen. In Pechlaner, H. and Glaeßer, D. (Eds.), *Risiko und Gefahr im Tourismus. Erfolgreicher Umgang mit Krisen und Strukturbrüchen (pp. 131–141)*, Berlin, Germany: Erich Schmidt Verlag.

Spiegel online (2017, January 16). *Angst vor hartem Brexit lässt Pfund fallen*. Retrieved from http://www.spiegel.de/wirtschaft/soziales/grossbritannien-angst-vor-hartem-brexit-laesst-pfund-fallen-a-1130112.html.

Sönmez, S. F. (1998). Tourism, Terrorism, and Political Instability. *Annals of Tourism Research, 2(2)*, 416–456.

Sönmez, S. F., Apostolopoulos, Y. and Tarlow, P. (1999). Tourism in Crisis: Managing the Effects of Terrorism. *Journal of Travel Research, 38(1).*, 13–18.

Statistisches Bundesamt (Ed.) (2013–2017). *Übernachtungen aus dem Quellmarkt "Vereinigtes Königreich"*. Retrieved from online database. Wiesbaden, Germany.

Statistical office of the European Union (Eurostat) (Ed.) (2017a). *Number of trips by country / world region of destination [tour_dem_ttw]. [xls]*. Retrieved from online database: http://ec.europa.eu/eurostat/data/database.

Statistical office of the European Union (Eurostat) (Ed.) (2017b). *Number of nights spent by country / world region of destination [tour_dem_tnw]. [xls]* Retrieved from online database: http://ec.europa.eu/eurostat/data/database.

Steiner, Chr., Al-Hamarneh, A. and Meyer, G. (2006). Krisen, Kriege, Katastrophen und ihre Auswirkungen auf den Tourismusmarkt. *Zeitschrift für Wirtschaftsgeographie, 50(2),* 98–108.

Süddeutsche Zeitung (2016, February 20). *Hintergrund: Die Versprechen der EU an Großbritannien.* Retrieved from: http://www.sueddeutsche.de/news/politik/eu-hintergrunddie-versprechen-der-eu-an-grossbritannien-dpa.urn-newsml-dpa-com-20090101-160220-99-884405.

Theurer, M. (2016a, February 23). *Worüber die Briten wirklich abstimmen.* Frankfurter Allgemeine Zeitung (FAZ) Retrieved from: http://www.faz.net/aktuell/wirtschaft/wirtschaftspolitik/brexit-hintergruende-zum-eu-referendum-von-david-cameron-14048572.html.

Theurer, M. (2016b, June 13). *Brexit-Angst lässt den Pfundkurs taumeln.* Frankfurter Allgemeine Zeitung (FAZ). Retrieved from: http://www.faz.net/aktuell/finanzen/devisen-rohstoffe/investoren-stossen-pfund-aus-angst-vor-brexit-ab-14285451.html.

Travel Industry Club & manufacts research & dialog (Ed.) (2016). *Entscheiderpanel der Reiseindustrie - Ergebnisse Juli 2016.* Retrieved from: https://www.travelindustryclub.de/brexit-wird-sich-nachhaltig-auf-den-tourismus-auswirken-eine-umfrage-des-travel-industry-club/.

Zeit online (2016, October 10). *Britisches Pfund fällt auf niedrigsten Stand seit 30 Jahren.* Retrieved from: http://www.zeit.de/wirtschaft/2016-10/brexit-britisches-pfund-fall-tiefstand-grossbritannien.

List of Contributors

Dr Alisha Ali (PhD., MSc., BSc., FHEA) is a Principal Lecturer and Sheffield Business School's PhD Programme Leader at Sheffield Hallam University. Her research interests are sustainable development, technology applications and impacts for hospitality and tourism, innovation in hospitality and tourism, destination management, applied learning and online research methods. She teaches in the areas of innovation, IT, sustainable tourism and hospitality operations. In addition to academic activities, Alisha has worked both locally, regionally and internationally in the hospitality and tourism field. Some of her career highlights includes developing marketing communication material for the travel and tourism industry, conducting market research and analysis of world-wide trends in the hospitality industry, developing feasibility studies for industry stakeholders and working on live tourism plans and projects and participating in their implementation in the Caribbean. Alisha is building a research domain focusing on ICT and sustainable tourism and guest edited special issues of the *Journal of Information Technology & Tourism* on ICT for sustainable tourism and ran workshops in this area. Alisha also has a background in consultancy working with government offices, destination management organisations (DMOs) and hospitality and tourism businesses.

Dr Elricke Botha is a senior lecturer in tourism management in the Department of Applied Management at the University of South Africa, where she has been a staff member for approximately 10 years. Apart from her teaching experience, Elricke also has managerial experience as the Coordinator: Postgraduate Studies in the College of Economic and Management Sciences and have served on several teaching and research related committees in this regard. Elricke completed her PhD in Tourism Management at the North-West University, South Africa where she developed an interpretation framework for the Kruger National Park, South Africa. Since then Elricke's publications has mainly focused on sustainable tourism management and related topics. ORCID iD: 0000-0003-1761-8264

Yasmine M. Elmahdy is a PhD candidate in the School of Sport and Recreation at Auckland University of Technology, New Zealand. Her current research focuses on sustainable management of marine mammal tourism in New Zealand. Her research interests lie in the areas of marine tourism, ecotourism, wildlife tourism, adventure tourism, and extreme and adventure sports. Her other research interests include gender issues in tourism and sport, and qualitative approaches to research.

Dr Marit Gundersen Engeset is professor of marketing at University of South-Eastern Norway. She is the programme director for the PhD programme in Marketing Management at the University of South Eastern Norway. During her 25 years of experience with tourism and consumer research she has been focusing on areas such as guest value, tourist satisfaction, travel motives, and creativity. Her work is published in international journals, such as *Journal of Travel Research, Cornell Quarterly, International Journal of Hospitality Management, Tourism Review, International Journal of Culture, Tourism, and Hospitality Management, Journal of Marketing Research,* and *Journal of Consumer Marketing*.

Eva Holmberg, Lic. Econ, is senior lecturer of tourism and methods HAAGA-HELIA University of Applied Sciences. She teaches students at both bachelor and master in courses such as brand management, responsible tourism and destination management. Her research interests are mainly related to destination management, branding and responsible tourism development and inquiry learning.

Amber Knowsley was a Master of International Hospitality Management student at Auckland University of Technology, graduating with Honours in 2017. Further to her research on carbon mitigation in the New Zealand accommodation industry, she is interested in how businesses in other areas of the hospitality industry can improve guest experience and satisfaction, ideally within a sustainable framework. She is currently providing consulting services to organisations in Auckland, New Zealand.

Dipl.-Kffr. Anne Köchling (MTM) studied International Business Studies at the University of Paderborn (Germany) and the University of Santiago de Compostela (Spain). After finishing her degree in Business Administration she complemented a Master of Tourism Management at the « Freie Universität » in Berlin (Germany). In the year 2005 she started her professional career in destination management at the German National Tourist Board in Frankfurt (Germany). From 2007 until 2009 she worked at the the tourism organisation for the federal state of Schleswig-Holstein where she implemented the marketing for the destination. Since 2010 she has been working at the DITF - German Institute for Tourism Research (formerly known as IMT) at the West Coast University of Applied Sciences in Heide/Holstein (Germany), where she is currently in charge of several research projects related to the field of destination management and marketing. Furthermore, she is pursuing her doctorate at the Leuphana University in Lüneburg, where she focuses on the destination experience on destination websites.

Dr Claire Liu is Head of Department - Tourism and Events and Senior Lecturer in the School of Hospitality and Tourism, Auckland University of Technology in New Zealand. Claire holds a PhD in Tourism Management from Massey University, New Zealand. Her research areas include sustainable tourism management, SMEs and tourism entrepreneurship, Chinese outbound tourist behaviours, and tourism and hospitality education. Claire serves on the editorial boards for *International Journal of Tourism Cities* and *Journal of China Tourism Research*. She is the Executive Representative of Australasia for the International Tourism Studies Association (ITSA). In 2019, Claire became the New Zealand country consultant for APacCHRIE. Her recently published book *Tourism Education and Asia* (Liu & Schänzel, 2019) has discussed the current status, pertinent issues, and challenges in tourism education which provide particular interest to Western educators as a window to the Asian tourism education landscape. ORCID iD 0000-0002-2533-749X

Dr Michael Lück is a professor in the School of Hospitality and Tourism at Auckland University of Technology, New Zealand. He is founding co-chair of the International Coastal & Marine Tourism Society (ICMTS). Michael has more than 10 years' work experience in the tourism industry and his research interests include (marine) wildlife tourism, the cruise industry, ecotourism, interpretation and education on wildlife tours, the impacts of tourism, and aviation. He has published in a number of international journals, is founding editor of the academic journal *Tourism in Marine Environments*, Associate Editor of the *Journal of Ecotourism,* and editorial board member of *Marine Policy*. Michael has edited or co-edited more than ten volumes on ecotourism, marine and polar tourism, events and low cost airlines, as well as the *Encyclopedia of Tourism and Recreation in Marine Environments* (CABI), and co-authored the introductory text *Tourism* (2nd. Ed., CAB International). ORCID iD: 0000-0002-6473-8579

Dr Philip Murray is a Senior Teaching Fellow in Operations Management in the Department of Business Transformation at Surrey Business School, where he teaches and supervises students on undergraduate and postgraduate programmes. Philip is a Fellow of the HEA and an experienced programme and module manager who has an active role in curriculum and new programme development. Philip has also been active in preparing programmes for academic quality review and external validation on several occasions both in Ireland and the UK. He has previously worked as a Senior Lecturer and Course Leader in the Department of Service Sector Management at Sheffield Hallam University (2016–2019) and as a Lecturer at Cork Institute of Technology (2004 – 2016). Prior to joining the academy Philip worked in operations for international hotel brands, such as Crowne Plaza and Clarion Hotels, as well as SME's in the tourism sector.

Donna O'Donnell is a tourism lecturer at Auckland University of Technology. Her research centres around two major themes: Education and tourism. Donna's research interests include educational tourism, tourism education, technology and learning, and dealing with Dyslexia in an academic environment. She is passionate about the tourism industry, teaching and learning and enhancing the individual learning experience.

Professor Mark Orams is the Dean of the Graduate Research School and Interim Dean of the Faculty of Health and Environmental Sciences at Auckland University of Technology, New Zealand. His interests lie in coastal and marine recreation, tourism and sport and he has published widely in these areas. He has a background as a professional yachtsman and has competed in a range of offshore sailing events as well as elite level small boat regattas. His expertise in competitive sailing is widely sought after in the media where he is known as The Sailing Professor. ORCID iD: 0000-0001-6806-7891

Dr Chantal Denise Pagel (M.Sc./ M.I.N.C.) has recently completed her PhD at Auckland University of Technology, New Zealand. She has been engaged in marine wildlife tourism research since 2010 and has specialised in commercial in- water interactions with marine wildlife, which she has further explored in her doctoral research project. ORCID ID: 0000-0002-2979-0806

Dr Tomas Pernecky is Professor in the School of Hospitality and Tourism, Faculty of Culture and Society at the Auckland University of Technology, New Zealand. He is passionate about investigating the ways in which social realities are constructed, and the implications and possibilities this denotes for different peoples and communities. His research interests are broad and multifaceted, ranging from the philosophy of science to specific areas of phenomenology, social constructionism, social ontology, post-existentialism, sustainable leadership and a host of conceptual, theoretical and methodological issues examined in the context of events, tourism, hospitality and leisure.

Associate Professor Jill Poulston (PhD) is semi-retired, and currently works as a contract doctoral supervisor. Immediately prior to this, she led the hospitality postgraduate programmes at Auckland University of Technology where she was also an Associate Director of the New Zealand Tourism Research Institute. She has published her work on a variety of ethical issues and hospitality workplace problems, as well as related topics such as organic food, sustainability, sexual harassment, and diversity. Jill most recently taught leadership and ethics to postgraduate students, examines master's and doctoral research on her topics of interest, and continues to publish her research.

Dipl.-Geogr. Julian Reif graduated as geographer at the University of Bonn and has been project manager at the DITF - German Institute for Tourism Research (formerly known as IMT) at the West Coast University of Applied Sciences since 2012. Julian has several years of experience in the management and implementation of primary and secondary market research projects for tourism demand. His main areas of work and research interests are: Tourism demand, digitalisation in tourism and tourism impacts. He is currently pursuing his doctorate at the University of Bonn on the measurement of tourist spatio-temporal behavior.

Jarmo Ritalahti (Lic.Phil.) is a principal lecturer of tourism at Haaga-Helia University of Applied Sciences and Head of the Master Degree in Aviation and Tourism Business. His main areas of expertise are tourism, travel intermediation and pedagogical and curriculum development in higher education. Jarmo has worked on numerous development and research projects, especially in pedagogical development and tourism as a regional development tool. His roles in the projects have varied from a project researcher to project manager, member, and chair of steering groups. He has published more than ten peer-reviewed journal and reader articles, conference papers and educative books.

Professor Melville Saayman (PhD) was director of the research unit TREES (Tourism Research in Economic Environs and Society) formerly known as the Institute for Tourism and Leisure Studies, at the North-West University (Potchefstroom Campus) in South Africa. Until March 2009, he had also been head of the tourism programme at the same university for 17 years. He served on several boards as a director, including the South African Tourism Board (SATOUR), North-West Parks and Tourism Board, Institute of Environment and Recreation Management, National Zoological Council, South African National Recreation Council (SANREC), North-West Recreation Council (PROREC-NW) and North West Development Corporation. He was director of Aardklop National Arts Festival. At an international level, he was a member of the executive committee of the Association of International Experts in Tourism (AIEST) and also served on the World Tourism Organisation's panel of experts. He became the first South African to be nominated as resource editor of the leading tourism journal, *Annals of Tourism Research*. He also served on various other editorial boards and has published in most of the major national and international tourism journals. He was active in the field of tourism and leisure economics and development with a clear focus on poverty alleviation. He has done significant research in the field of event economies, nature-based tourism, marine tourism and how these types of tourism alleviate poverty. He became the first National Research Foundation (NRF) rated researcher in tourism in South Africa. From his pen, numerous leisure and tourism books (20), scientific articles (240), technical reports (420) and in-service training manuals (8) have been published. He was study leader and promoter to 95 master's and doctoral students and presented more than 100 papers at international conferences. He has also been an examiner of more than 50 Masters and PhD theses. In 2010, 2011, 2014 and 2015 he was awarded researcher of the year of North-West University. In 2017 he was acknowledged by the Department of Science and Technology for his contribution to Sustainable Tourism. This made him the first South African who received an award from the Department of Science and Technology in the field of Tourism.

Dr Heike Schänzel is an Associate Professor and Programme Leader for the Master of International Tourism Management at Auckland University of Technology, Auckland, New Zealand. Her research interests include families, children, and gender in leisure and tourism; femininities and masculinities in tourism research; tourist experiences and social dimensions. She is passionate about better understanding family fun (along with the avoidance of conflict) and gender along with the facilitation of sociality and meaningful experiences within the context of leisure and tourism. ORCID iD: 0000-0002-3885-5640

Michael Scheer graduated as a biologist at the University of Bremen where he studied cetacean bioacoustics and behaviour. Michael's research focused on the behaviour of marine mammals during touristic wildlife encounters. He recently worked as an independent scholar for the Institute for Terrestrial and Aquatic Wildlife Research (ITAW) and the Department of Agriculture, Environment and Rural Areas (LLUR) of Schleswig-Holstein on the management of grey seals and tourism. ORCID ID: 0000-0001-5244-5042

Dr Sabrina Seeler is a lecturer in International Tourism Management at the West Coast University of Applied Sciences, Heide (Germany). Sabrina holds a PhD from Auckland University of Technology, Auckland (New Zealand) and has worked as a postdoctoral researcher in experience-based tourism at Nord University Business School, Bodø (Norway). Being a critical realist and having experiences in qualitative and quantitative research methods, her research interests include consumer behaviour in tourism; experience consumption and creation; transformative experiences; strategic destination management; and sustainable visitor management.

Dr Peet van der Merwe is currently a professor and researcher at the School of Tourism Management, at the Faculty of Economic Management Sciences. He also forms part of the research unit TREES (Tourism Research in Economics, Environs and Society), at the faculty which is the only research unit in South Africa with a focus on tourism relates aspects. His main area of specialisation lies with natural-area tourism, which includes ecotourism, wildlife tourism, adventure tourism, marine ecotourism and sustainable tourism development. Prof van der Merwe has done numerous research project for different organisations such as South African National Parks, North West Parks Board, South African Predator Association (SAPA), Wildlife Ranching South Africa, Namibia Wildlife Ranching, SA Hunters, Confederation of Hunters Associations of South Africa (CHASA), Professional Hunters Association of South Africa (PHASA), Northern Cape Tourism and Anglo Gold. Prof van der Merwe is national and international known for his research on consumptive and non-consumptive wildlife tourism. Prof van der Merwe is an NRF-rated researcher (National Research Foundation) with various research outputs: Peer-reviewed publications (65), Popular articles (44), Projects (48), Research reports (38), Books (4), Chapters in books (5), post graduate students (PhD and Masters - 27), conferences attendance (International and National - 60) and International guest lectures.

Dr Jan Velvin is Head of department of Business, Strategy and Political Sciences at USN Business School, University of South Eastern Norway. He also holds a position as Director of the Executive master management programme and the bachelor programmes for part time management studies at USN Business School. Jan has worked on issues in tourism since 1994. His research interests are rural areas development, embedded innovation, service management, and second home tourism.

Dr Matthias Walter is professor and the leader of a wildlife conservation workgroup at the University of Goettingen, Germany, and director of the bi-national MSc programme *International Nature Conservation* with Lincoln University, New Zealand. He has a focus on terrestrial vertebrates from the old world tropics and the Middle East but is also passionate about whale watching.

Rebekka Weis (M.A.) studied International Tourism Management at the West Coast University of Applied Sciences, Heide (Germany). Since 2012 she has worked as a market researcher at the DITF - German Institute for Tourism Research (formerly known as IMT) at the West Coast University of Applied Sciences, Heide (Germany). Her areas of interest include quantitative market research methods for tourism, nature and adventure tourism as well as horse-based tourism.

Schriftenreihe des Instituts für Management und Tourismus (IMT)

Herausgegeben von der Fachhochschule Westküste

Die Bände 1-6 sind im Martin Meidenbauer Verlag erschienen und können über den Verlag Peter Lang, Internationaler Verlag der Wissenschaften, bezogen werden: www.peterlang.com.

Ab Band 7 erscheint diese Reihe im Verlag Peter Lang, Internationaler Verlag der Wissenschaften, Frankfurt am Main.

Band 7 Anja Wollesen: Die Balanced Scorecard als Instrument der strategischen Steuerung und Qualitätsentwicklung von Museen. Ein Methodentest, unter besonderer Berücksichtigung der Anforderungen an zeitgemäße Freizeit- und Tourismuseinrichtungen. 2012.

Band 8 Wolfgang Georg Arlt (Ed.): COTRI Yearbook 2012. 2012.

Band 9 Michael Lück / Jan Velvin / Bernd Eisenstein (eds.): The Social Side of Tourism: The Interface between Tourism, Society, and the Environment. Answers to Global Questions from the International Competence Network of Tourism Research and Education (ICNT). 2015.

Band 10 Bernd Eisenstein / Christian Eilzer / Manfred Dörr (Hrsg.): Kooperation im Destinationsmanagement: Erfolgsfaktoren, Hemmschwellen, Beispiele. Ergebnisse der 1. Deidesheimer Gespräche zur Tourismuswissenschaft. 2015.

Band 11 Michael Lück / Jarmo Ritalahti / Alexander Scherer (eds.): International Perspectives on Destination Management and Tourist Experiences. Insights from the International Competence Network of Tourism Research and Education (ICNT). 2016.

Band 12 Lars Rettig: Digitalisierung der Bildung. Warum und wie lernen wir ein Leben lang? Forschungsergebnisse zur Online-Weiterbildung im Tourismus. Bedeutung – Erwartung – Nutzung. 2017.

Band 13 Bernd Eisenstein / Christian Eilzer / Manfred Dörr (Hrsg.): Demografischer Wandel und Barrierefreiheit im Tourismus: Einsichten und Entwicklungen. Ergebnisse der 2. Deidesheimer Gespräche zur Tourismuswissenschaft. 2017.

Band 14 Alisha Ali / John S. Hull (eds.): Multi-Stakeholder Perspectives of the Tourism Experience. Responses from the International Competence Network of Tourism Research and Education (ICNT). 2018.

Band 15 Anja Wollesen / Christian Eilzer / Manfred Dörr (Hrsg.): Nachhaltigkeit im Tourismus unter besonderer Berücksichtigung von kleinen Tourismusgemeinden: Herausforderungen, Implementierung, Monitoring. Ergebnisse der 3. Deidesheimer Gespräche zur Tourismuswissenschaft. 2020.

Band 16 Dirk Schmücker / Eric Horster / Edgar Kreilkamp: Digitalisierung – Chance oder Risiko für nachhaltigen Tourismus? Eine Studie im Auftrag des Umweltbundesamtes (UBA) zu den Auswirkungen von Digitalisierung und Big-Data-Analyse auf eine nachhaltige Entwicklung des Tourismus und dessen Umweltwirkung. 2020.

Band 17 Michael Lück / Claire Liu (eds.): A kaleidoscope of tourism research: Insights from the International Competence Network of Tourism Research and Education (ICNT). 2021.

www.peterlang.com

www.ingramcontent.com/pod-product-compliance
Ingram Content Group UK Ltd.
Pitfield, Milton Keynes, MK11 3LW, UK
UKHW021842210426
5322IPUK00022B/417